ARATINGAS

HERMAN KREMER

CIP-GEGEVENS KONINKLIJKE BIBLIOTHEEK, DEN HAAG

Kremer, Herman

Aratingas/Herman Kremer; [photogr.: Cees Scholtz;
ill.: Bauke Vliegendehond; transl. from the Dutch].-
Noordbergum: "Ornis". - Ill., foto's
Vert. van: Aratinga's. Noordbergum: Ornis, 1989.
Met lit. opg., reg.
ISBN 90-73217-04-0 geb.
Trefw.: Aratinga's.

Photographs: Cees Scholtz

Illustrations: Bauke Vliegendehond

Front cover: Golden-headed Conure

October 1992

Distributed in the United Kingdom by
The Owl's Nest Bookshop, Birdworld,
Holtpound, Nr Farnham Surrey GU10 4LD
Tel. (0420)22668 Fax (0420)23715

Distributed in all other countries by
Uitgeverij "Ornis"
de Zwette 58
NL-9257 RR Noordbergum
the Netherlands
Tel. 05110-73589 Fax 05110-73549

ISBN 90-73217-04-0

Copyright © 1992 Herman Kremer

Published by:
Uitgeverij "Ornis"
NL-9257 RR Noordbergum

All rights reserved. No part of this book may be reproduced or transmitted in any form or by any means, electronical or mechanical, including photocopying, recording or any information storage and retrieval system, without permission in writing from the Publisher.

CONTENTS

I.	Introduction	11
	General characteristics	13
	Distribution	13
	Species and sub-species	15
II.	Purchasing	17
III.	Habits and sex	20
IV.	Accomodation	24
V.	Diet	27
VI.	Breeding	30
VII.	Diseases	35
	Measures and requirements	36
VIII.	Administration	38
IX.	Species descriptions	40
	Textual divisions	41
	Sharp-tailed Conure	45
	Blue-crowned Conure	45
	Margarita Conure	45
	Bolivia Conure	45
	Queen of Bavaria's Conure	51
	Mexican Green Conure	55
	Socorro Green Conure	55
	Brewster's Green Conure	55
	Nicaraguan Green Conure	55
	Red-throated Conure	55
	Finsch's Conure	61

Wagler's Conure	65
Peter's Conure	65
Cordilleras Conure	65
Carriker's Conure	65
Mitred Conure	71
Chapman's Mitred Conure	71
Red-masked Conure	77
White-eyed Conure	83
Ecuadorian White-eyed Conure	83
White-eyed Conure (*A. l. propinquus*)	83
(Nicéforo's) White-eyed Conure	83
Hispaniolan Conure	87
Mauge's Conure	87
Cuban Conure	93
Golden-headed Conure	97
Golden-fronted Conure	97
Jendaya Conure	103
Sun Conure	107
Weddell's Conure	113
Jamaican Conure	119
Aztec Conure	119
Eastern Aztec Conure	119
Petz's Conure	123
Western Mexican Petz's Conure	123
Petz's Conure (*A. c. clarae*)	123
St Thomas Conure	131
Bonaire Brown-throated Conure	131
Aruba Brown-throated Conure	131
Brown-throated Conure	131
Colombia Brown-throated Conure	131
Lehmann's Brown-throated Conure	131
Tortuga Brown-throated Conure	131
Margarita Brown-throated Conure	131
Venezuela Brown-throated Conure	131
Guiana Brown-throated Conure	131
Surinam Brown-throated Conure	131
Golden-cheeked Brown-throated Conure	131

Para Brown-throated Conure	131
Brown-eared Conure	131
Cactus Conure	141
Pale Cactus Conure	141
Golden-crowned Conure	145
Greater Golden-crowned Conure	145

Acknowledgements

Literature

Register

I. Introduction

The world of the parrots and parrot-like birds is characterized by its extensiveness and its great diversity. Worldwide there are no fewer than 344 species or thereabouts, and on top of them a further 445 sub-species (a number of which have incidentally already become extinct). They inhabit Australia and the surrounding islands, the whole of southern Asia, Africa, and South and Central America. Their distribution range used to be even larger: fossil remains of parrotlikes have even been found in Europe, and the last Carolina Parakeet in North America probably died in the mid nineteen twenties.

This book deals with a comparitively small group of parrot-likes, namely the birds belonging to the genus *Aratinga*. They are found together with members of other genera from South and Central America. According to Forshaw this region contains a total of 27 genera, 141 species and a further 156 sub-species. A number of these are familiar to birdkeepers, for example the genera *Ara, Aratinga, Pyrrhura, Brotogeris, Forpus, Pionus* and *Amazona*.

Although they have yet to achieve the popularity of the Australian parakeets, there is a growing number of enthusiasts who see it as a challenge to try their hand at keeping these South American birds. The first problem which often arises is a certain unfamiliarity and a lack of knowledge of the possibilities: which species are available, what they cost, how are they bred, can they be treated in the same way as Australians?, etc. etc.
Articles are occasionally published in magazines but there is very little good and systematic information available on these points, at least not in the sense that sufficient practical information about each species is given to the birdkeeper.

Partly due to the exceptionally warm reception which my book "Australian Parakeets and their Mutations" received, and the resulting demand for something similar on South American parakeets, I have attempted in this book to make a start on the subject. The first question was, which birds should the book be about? All the South Americans or only a number of them? In the first instance I have deliberately limited myself to one group, the genus *Aratinga*. The arguments for this restriction are as follows:

a. a single book about all the South American parrots would require many years of preparation;
b. such a book would be an expensive publication, and it is my particular aim to produce something which is available to every aviculturist;
c. not all birdkeepers are interested in all the South and Central American parrotlikes. By tackling each group separately everybody can concentrate on his own preference.

Besides the description of each species a number of chapters on purchase, sexing, housing, breeding, diseases, and so on have also been included. However, these have been kept short, firstly because more general descriptions are regularly dealt with in various

1. In Ecuador the occasional Red-masked Conure can still be encountered high in the Andes.

other books and magazines, and secondly because these aspects are sometimes gone into more deeply in the separate species' descriptions.

General characteristics

Scientists divide the genus *Aratinga* into species and sub-species in a variety of ways. The number of species varies from between 15 and 21, and the number of sub-species from 53 to 57. In this book the classification made by Forshaw in the second edition of "Parrots of the World" is used, which means that we are concerned with 19 species and, in addition, a total of 35 sub-species (one of which is extinct). There are seven species without any sub-species, whereas the St Thomas Conure has no fewer than 13 sub-species besides the nominate form.

Rather a lot of variation in colour and size occurs within the genus *Aratinga*, which means that really every birdkeeper is able to find something to suit their taste.
The most beautiful representative of this genus is the Queen or Bavaria's Conure which, with its brilliant golden yellow colouring, is a feast for the eye. Unfortunately this very bird is also one of the rarest, and certainly the most expensive.
On the other hand there are a number of green to brown-green (sub-)species which are less striking, such as the White-eyed, the Hispaniolan, the Jamaican and several of the Brown-throated Conures. Of course each of these (sub-)species nevertheless also has its own appeal.
Finally there is a considerable variation in size: from the Petz's, the Jamaican and several of the Brown-throated Conures at 24cm (9.5in) to the Cordilleras Conure at 42cm (16.5in). The members of this genus are, to a greater or lesser degree, similar in appearance to the macaws; the resemblance is particularly striking when one of the dwarf macaws is placed next to, for example, a Sharp-tailed Conure.

Many birdkeepers have difficulty in identifying the various species and sub-species. For this reason I have tried as far as possible to include a good photograph of each one in the book, as this is, after all, the best way to reproduce the birds' colouring. Unfortunately not all the sub-species are available in the Netherlands, so it has not been possible to include them all. However, the differences between the sub-species are indicated as clearly as possible within the descriptions of the species under which they fall.

Distribution

The accompanying map shows the complete distribution range of the 19 species of Aratingas. It will be noticed that they are found in a large part of South and Central America, from Mexico to central Argentina. They also occur on various islands in the Caribbean. The term "South American parrots" is therefore not entirely correct.
Several striking aspects of the distribution are the following:
- Aratingas inhabit the northern part on the Andes Range, but not the southern part (between Chile and Argentina). Other parrot-likes are found in this latter area;
- in a long strip in the south-east of South America the range does not reach the coast;

- there is a strange and illogical gap in the range, as Aratingas are totally absent in a large part of Panama;
- each island has its own species or sub-species.

The habitat of these birds in this immense distribution range is extremely varied. It includes tropical rainforest, savanna, and desert or semidesert, sometimes in combination with mountains. Compared to the situation in Western Europe the temperatures are fairly constant throughout the whole year, as are the number of hours of daylight. This accounts for the fact that they are not especially restricted to one time of the year when it comes to breeding. In Europe they lay eggs from early in spring to late in summer. However, in general they are later than Australian parakeets.

Species and sub-species

As a preface to the descriptions per species the following is a full survey of all the species and sub-species within this genus. Both the scientific and the English names are mentioned. If there are sub-species, and the scientific name therefore consists of three words, in the nominate form the second and the third are always the same.

1. *Aratinga acuticaudata acuticaudata* - Sharp-tailed Conure
   Aratinga acuticaudata haemorrhous - Blue-crowned Conure
   Aratinga acuticaudata neoxena - Margarita Conure
   Aratinga acuticaudata neumanni - Bolivia Conure

2. *Aratinga guarouba* - Queen of Bavaria's Conure

3. *Aratinga holochlora holochlora* - Mexican Green Conure
   Aratinga holochlora brevipes - Socorro Green Conure
   Aratinga holochlora brewsteri - Brewster's Green Conure
   Aratinga holochlora strenua - Nicaraguan Green Conure
   Aratinga holochlora rubritorquis - Red-throated Conure

4. *Aratinga finschi* - Finsch's Conure

5. *Aratinga wagleri wagleri* - Wagler's Conure
   Aratinga wagleri transilis - Peter's Conure
   Aratinga wagleri frontata - Cordilleras Conure
   Aratinga wagleri minor - Carriker's Conure

6. *Aratinga mitrata mitrata* - Mitred Conure
   Aratinga mitrata alticola - Chapman's Mitred Conure

7. *Aratinga erythrogenys* - Red-masked Conure

8. *Aratinga leucophthalmus leucophthalmus* - White-eyed Conure
   Aratinga leucophthalmus callogenys - Ecuadorian White-eyed Conure
   Aratinga leucophthalmus propinquus - White-eyed Conure
   Aratinga leucophthalmus nicefori - (Nicéforo's) White-eyed Conure

9.	*Aratinga chloroptera chloroptera*	- Hispaniolan Conure
	Aratinga chloroptera maugei	- Mauge's Conure
10.	*Aratinga euops*	- Cuban Conure
11.	*Aratinga auricapilla auricapilla*	- Golden-headed Conure
	Aratinga auricapilla aurifrons	- Golden-fronted Conure
12.	*Aratinga jandaya*	- Jendaya Conure
13.	*Aratinga solstitialis*	- Sun Conure
14.	*Aratinga weddellii*	- Weddell's Conure
15.	*Aratinga nana nana*	- Jamaican Conure
	Aratinga nana astec	- Aztec Conure
	Aratinga nana vicinalis	- Eastern Aztec Conure
16.	*Aratinga canicularis canicularis*	- Petz's Conure
	Aratinga canicularis eburnirostrum	- Western Mexican Petz's Conure
	Aratinga canicularis clarae	- Petz's Conure
17.	*Aratinga pertinax pertinax*	- St Thomas Conure
	Aratinga pertinax xanthogenia	- Bonaire Brown-throated Conure
	Aratinga pertinax arubensis	- Aruba Brown-throated Conure
	Aratinga pertinax aeruginosa	- Brown-throated Conure
	Aratinga pertinax griseipecta	- Colombia Brown-throated Conure
	Aratinga pertinax lehmanni	- Lehmann's Brown-throated Conure
	Aratinga pertinax tortugensis	- Tortuga Brown-throated Conure
	Aratinga pertinax margaritensis	- Margarita Brown-throated Conure
	Aratinga pertinax venezuelae	- Venezuela Brown-throated Conure
	Aratinga pertinax chrysophrys	- Guiana Brown-throated Conure
	Aratinga pertinax surinama	- Surinam Brown-throated Conure
	Aratinga pertinax chrysogenys	- Golden-cheeked Brown-throated Conure
	Aratinga pertinax paraensis	- Para Brown-throatred Conure
	Aratinga pertinax ocularis	- Brown-eared Conure
18.	*Aratinga cactorum cactorum*	- Cactus Conure
	Aratinga cactorum caixana	- Pale Cactus Conure
19.	*Aratinga aurea aurea*	- Golden-crowned Conure
	Aratinga aurea major	- Greater Golden-crowned Conure

II. Purchasing

Aratingas are available from a number of birdkeepers, but some species are also still being imported. Particularly with newly-imported birds it is important to be on your guard, not only because the risk of disease is greater, but also because they still have to become accustomed to our climate and diet.
With regard to diet it is advisable for every birdkeeper to note what kind of food the bird has at the time of purchase. If possible they should be fed on a similar diet, even if it is not ideal, and then gradually put onto a better diet.

Newly-imported Aratingas are usually extremely shy, so to start with, they should left in peace as much as possible to allow them to get used to their new situation.
The most common complaints among newly-imported birds are intestinal disorders. These are often fairly easy to cure: warmth, extra vitamins, not too much moist feed and clean living conditions. If possible it is better not to use any antibiotics, as they will already have been pumped full of them during quarantine. Another course might have a bad effect, as antibiotics destroy not only the damaging elements in the intestinal flora, but also the useful ones. As a result they become extremely susceptible to all kinds of new infections. To be on the safe side a worming agent can be given.

Before a purchase is definitely concluded the bird being considered should be thoroughly examined. First of all you must take note of how the bird behaves in the aviary or cage in which it is housed, and preferably in such a way that it does not know that it is being studied. Only in these circumstances will it behave as naturally as possible. It should be alert and lively, and its plumage should have a smooth appearance. Do not be deceived by lively behaviour when you come closer as fear then alters the normal behaviour and it becomes impossible to draw any accurate conclusions. Further, you must take note of, among other things, the body, droppings, eyes, nostrils, plumage and legs. The body should feel warm to the touch, particularly the breast.
There is no objection to being able to feel the breastbone, but it should not stick out sharply. In other words the bird should be "round-breasted".

Also study the droppings in the cage or aviary in which the bird is being kept. They should not be too loose and watery, this indicates that something is wrong. If the bird is still in your travelling box or has just arrived in a strange aviary, loose droppings are fairly normal as the bird is then unsettled and nervous and this affects its condition. It will get better after a while. An extra check is to closely study the feathers around the vent, these should be clean and dry. If the droppings have been loose and watery for a longer period the feathers will be sticky and dirty.

The eyes can tell the experienced birdkeeper a great deal. They should be clear and open. If they give a permanently sleepy impression or are swollen there is a fair chance that something is wrong.

Generally speaking the plumage should be smooth and shiny, although with newly-imported birds this does not need to be a deciding factor. After all, these birds go through a great deal in a fairly short time: they are first caught, and then put in a cage with many others by the catcher and sent to the exporter, and subsequently shipped overseas, all together and often in poor circumstances, to their destination. There they are put, once again in large numbers, into the quarantine quarters, and then sent to the pet shops or dealers before finally reaching the birdkeeper. With this sort of immediate past it is no wonder that their appearance is not too good at that moment. However, if they otherwise make a healthy and lively impression this does not have to form any problem. After a time the birds will moult and then their smooth and tidy plumage will automatically re-appear.

The legs should be properly formed and strong. A missing claw is no cause for concern; it never raises breeding problems, and certainly not where Aratingas are concerned, as they mate while next to each other on the perch, which gives them a great deal more grip than when the cock has to keep its balance on top of the hen.
The nostrils should be nice and clean and not damp, and the breathing should be inaudible.

"True pairs" are often offered for sale; however, always be careful when purchasing and never blindly believe everything you hear. Many birdkeepers are convinced that they have a true pair, for example because the birds have mated or because there is a clear difference between them.
However, this remains a difficult area, and with Aratingas one can only speak of a true pair after the birds have been sexed and produced fertile eggs.
You should take even greater care if you see a "proven pair" or "breeding pair" being offered. These terms may only be applied to a pair of birds which have bred successfully and reared young. In all other cases these terms are not applicable.

Further, more and more advertisements are appearing containing the words "breeding cock" or "breeding hen". These terms may only be applied to birds which have already reared young.

However, they will have done this together with one particular partner, which is apparently no longer available. On the face of it the information is useful, but it gives absolutely no guarantee that the bird in question will immediately and successfully start breeding with a new mate.

Another vague term is "sexually mature". With regard to Aratingas very little is known about this aspect. It is generally thought that the smaller species are able to start reproducing after two years, and that with the larger species it may well take a year longer. However, this is not known with certainty for all species. As many young birds also immediately have the outward appearance of their parents it is rather difficult to make definite statements about this matter.

In summary I will explain the terms again in short:
Cock 1.0; hen 0.1: a bird of which only the sex is mentioned.
Breeding male/breeding hen: a bird which has reared young together with a mate of the same species, and whose sex is guaranteed.

1.1: a cock and a hen of which only the sex is guaranteed.
Proven pair/breeding pair: a pair which has already reared young, and whose sex is guaranteed.

III. Habits and sex

Aratingas are social birds by nature. In the wild they usually move around in smaller or larger groups. It follows that they are quite different in character to the Australian species. If two birds are housed together they always try and make physical contact; they sit next to each other, preen each other, and spend the night in the nest-box together, regardless of whether they are birds of the same sex or not. They hardly ever show aggression towards each other. This only occurs when a number of birds are housed in one aviary and a pair has formed which wishes to start breeding. If there is not sufficient space, other birds will be chased away and even attacked. In the wild the various pairs are known to separate themselves from the group during the breeding season.

These parrots can become quite tame. In America they are very commonly kept as pets, which are more and more often birds that have been reared by hand. The most popular in this respect is the Petz's Conure, which is there called the Halfmoon Parrot. However, they never learn to speak very well, and their piercing voice can be exceptionally unpleasant in an enclosed indoor area.

Their rather loud voice is one of the Aratingas drawbacks, although they generally do not use it much. They are usually only noisy in the morning and evening, after waking and before going to sleep, as these are the times that they have to fill their stomachs either after or before a long night and further, when they are startled by unexpected occurrences. As their voice has a less pleasant character than that of the Australian parakeets they have also gained a worse reputation. A bird kept as a pet indoors will be fairly quiet, but if birds are outside and others of the same species are being kept nearby they will undoubtably keep contact by screeching.

However, there can be a great difference between individuals in this respect; one bird can carry on the whole day (although this is virtually unheard of) and another never makes a squeak. It is therefore impossible to say exactly which species make a lot or little noise. It is just as with people: the one never shuts his mouth, and the other never opens it.
Experience has shown that birds with eggs or young are fairly quiet.

A second drawback is that it is extremely difficult to determine the sex of these birds. In all the species the differences between the sexes is either too slight or there is too much variation in colour (for example in the amount of red on the green species) to be able to make any reliable judgement.

Differences in the rounding of the head and in the width of the base of the beak can sometimes give an indication, but they are no guarantees. For example I have a pair of sexed *Aratinga wagleri frontata*, which display no differences at all on these points, and yet they are undoubtably cock and hen. In general the head of the cock should be broader and heavier, and that of the hen slightly smaller and rounder. The base of the cock's beak should also appear broader when the birds are viewed from the front. Finally, the head of the cock should become broader towards the back when viewed from this same angle.

Another possibility is to carry out the pelvic bone test, which is used particularly for the genus *Agapornis*, and is based on the theory that the distance between the pelvic bones indicate whether the bird is a cock or a hen. This method is, however, not one hundred per cent reliable, although it is well worth applying if you wish to buy unsexed birds. It is then advisable to pick out the birds which show the greatest number of differences in order to increase the possibility of getting a pair. The distance between the pelvic bones can well give some indication, but the problem is that it is not always the same in all birds of one species, at least not where the hen is concerned. The distance will be greatest when the hen is laying, but then this test is not necessary as the production of eggs is proof enough. However, the distance will often also become greater when a hen is in breeding condition. If the hen is not laying, or is a young bird the position of the bones will be little or no different from that of the cock. All in all there is a fair chance that birds with a wide gap are hens, but a small gap tells you very little; these could be either young birds, or cocks or hens in rest. However, it is advisable to increase your experience of this test whenever you make a purchase. If you have selected birds in this fashion, and you let them undergo an endoscopic examination later, you can always see if your judgement on the basis of the pelvic bone test was correct.

And for those who are not entirely familiar with the test: it concerns the two small bones just in front of the vent. In adult males they are sometimes almost touching, whereas with other birds it is almost possible to place a finger between them, and these could therefore be hens. It is also said that in cocks the bones are sharper and more pointed and in hens more rounded.
The best way to apply this method is to hold the bird breast upwards in one hand, and then to feel the distance between the bones with the index finger of the other hand.

Another fairly reliable manier is to judge the birds from their behaviour. In order to do this you need a group of at least six birds together, as only two will tend to show courtship behaviour even if they are both cocks or both hens. Birds of the same sex frequently mate, so do not be deceived. Statistically you only have a chance of one to two that they are indeed a pair (with two birds there are the following possibilities: cock-cock, cock-hen, hen-cock and hen-hen).
However, in a larger group the cocks and hens will indeed pair off; they then more or less separate themselves from the rest and are fairly easy to spot.

The surest way of course, remains the endoscopic examination. However, not every birdkeeper is prepared to pay for this when birds worth only a few ponds are concerned. Of course everybody must decide for himself, but you can purchase a number of birds in this price range for the same amount, so the above-mentioned selection in a group makes most sense. Once a pair has been selected the rest of the birds can be resold.
The great advantage of an endoscopic examination is that it gives complete certainty and it also gives the vet a chance to examine other aspects: whether the bird is healthy, sexually mature, in breeding condition etc.

Such an endoscopic examination is possible after the bird is at least nine months old. Before this age the sexual organs are still insufficiently developed and virtually impossible to find.

2. Many small streams rise in the Andes, most of which flow into tributaries of the Amazon.

In summary, if you wish to attempt to select a pair by eye you should pay attention to the following aspects:
- in cocks the pelvic bones almost touch; in hens there is sometimes a larger gap, particularly when they are in breeding condition;
- the cock's head is broader and heavier;
- the base of the cock's beak is broader;
- the cock's head grows broader towards the back.

But once again, these differences never give complete certainty, and you are running somewhat of a risk if you rely on them.

Although there is not much information on the subject, most Aratingas will become sexually mature at two years old. This is after all when they are virtually fully coloured, and there is a strong relationship between sexual maturity and a full adult plumage. Consider, for instance, the Australian parakeets, such as the Red-winged Parakeet or the Pennant's Parakeet. Admittedly the latter does sometimes breed when one year old, but that is the result of being reared in captivity for years. In the wild they only breed in the second year.

IV. Accomodation

Aratingas are lively birds which are active throughout the greater part of the day, and therefore require the appropriate amount of space. The most suitable accomodation is that which allows for pairs to be housed in separate outdoor flights with adjoining lockable night shelters. In this way they can fully enjoy the natural conditions, which particularly in summer is very important, as then they can gain optimum benefit from the advantages of sun and rain.

The smaller species are sometimes kept in cages, but unless it is strictly necessary this is not recommended. If they spend all their time in a small cage their natural instinct for movement is suppressed and there is a chance that they will lose their natural lively character, become passive and even start feather-plucking. However, information from the United States where these birds are regularly kept and bred in cages sometimes no larger than 40 by 40 by 120cm, proves that this does not always have to be the case.

When you build aviaries for the smaller species you should think of a length of at least two metres (six feet), and for the largest species this should be four to five metres (twelve to fifteen feet).

In the main, the same construction can be used as for the Australian parakeets. However, take into account the fact that more of the South Americans are keen chewers, and that a wooden construction can become severely damaged. Consider, therefore, making them from metal. Generally speaking these will be more expensive than aviaries made of wood, but it will certainly be cheaper in the long run. What is more, a small-scale birdkeeper should be able to find a solution: for instance, rejected batches of unused metal which is suitable for our purposes can sometimes be bought cheap from scrap merchants.

As mentioned above, an outside flight with an adjoining lockable night shelter is ideal. Particularly in the cold and damp autumn and winter period it is important that the birds can be shut up to a night shelter which, if possible, can be heated. Although this is not strictly necessary, it does have its advantages: firstly Aratingas are happier in slightly higher temperatures, and secondly, heating keeps the air humidity down. This will undoubtably have a favourable effect on the condition of your birds as they are affected by cold and damp weather. It is therefore advisable to keep the entrance into the night shelter from the outside flight as small as possible. This hole can be placed in a door between the flight and the shelter, which can easily be left completely open in summer. Of course it must not become too warm in the night shelter in winter, because the difference with the temperature outside will become too great. This can result in a disturbance in the birds' natural biological rhythms, which among other things can lead to one of a pair going into an unwelcome moult so that they do not come into breeding condition at the same time.

The perches should not be too small; the toes should not be able to go right round them, as this will allow the claws to grow, there is then a much greater chance of them freezing in winter. For the smallest species they should be at least two and a half centimetres in diameter. Fresh (willow) branches have the advantage that the birds also tuck into them immediately, thus supplying them with some diversion and probably also certain nutrients. A disadvantage is that they have to be replaced more often. Perches made from hardwood last longest but will offer your birds little amusement.

In the wild many Aratingas live in the edges of the woods, or in other words in the transitional zone between wooded and open country. This is in fact a very common phenomenon in nature: the borders between habitats are the richest in wildlife.
Woods supply shelter and give the birds a feeling of security, therefore the birdkeeper should consider planting vegetation around the aviaries in such a way that his parrots get this same feeling. This will put them at their ease and undoubtably have a positive influence on their inclination to reproduce. Among other possibilities in this respect are either to have trees growing over the aviaries or to plant (non-poisonous) climbing plants up the wire netting of the flights. An additional advantage is that this improves the appearance of the aviaries. Greenery within Aratingas' aviaries does not usually survive for long; they like to chew living vegetation, and the plants' growth rate will not keep up with their destruction rate.

Finally you can also consider giving your Aratingas their freedom. In principle they are suitable for this as once they are used to their aviary and its surroundings, they will, normally speaking, always return to it. If you consider doing this - and what could be more beautiful than to see these fast fliers shooting free through the air - only begin once your birds are thoroughly used to their aviary. Also, to start with, only one of a pair should be let loose at a time, sometimes one, and sometimes the other. Besides its own aviary the emotional tie will also ensure that the fugitive returns. I hardly need to mention that there should of course always be sufficient food present in the aviary. Everyone would agree that this is the most ideal way of keeping birds. However, you must realize that the risks to the birds are much greater in this situation: bad weather, strong winds, predators, trigger-happy neighbours, etc., and that this would be against the law in the U.K.

There are undoubtably also birdkeepers who feel keen to keep Aratingas in colonies. I have tried it myself, but did not find it a success. I placed several Jendayas and Sharp-tailed Conures together in a single aviary with the following dimensions: a night shelter measuring three metres (9ft 9in) wide and one metre eighty centimetres (6ft) deep, divided into three compartments each measuring one metre (3ft 3in) by one metre eighty centimetres (6ft); these three opened into a flight of seven by three metres (23ft by 9ft 9in). The nest-boxes were hung in the shelter.
After a while the Sharp-taileds started breeding. All went well until the Jendaya Conures ventured into the compartment in the shelter where the Sharp-taileds' box was hung, and then all hell broke loose. The breeding pair aggressively attacked the Jendayas, and I was only able to save them from certain death as I happened to be in the neighbourhood at that moment: my attention was drawn by the horrendous screams.
Since then I have not attempted any similar experiments. However, with hindsight I sometimes think that things would have gone well if I had used a single large shelter instead of three smaller ones. This would have given the birds more chance to escape. The Sharp-taileds clearly considered the one compartment as exclusively their territory.

However, I have temporarily placed a number of birds together in one aviary several times. These were the occasions when I had purchased a number of imported birds at one time, and allowed them to pair themselves off. However, I found that also in this situation, once a pair had started breeding they became more and more aggressive towards the other birds, which then had to be removed as otherwise this would eventually have resulted in casualties.

It is advisable to house newly-bought Aratingas separately for the first few weeks, particularly if they are imported birds. If there is anything the matter with them this is the best way to do something about it. Problems are rarely serious, but you only once need to introduce an infectious disease and what you have built up over a period of years could be lost in one fell swoop. The greatest problem is namely that diseases are usually only discovered after the damage has already been done.

Another reason for supplying separate accommodation is when a bird has to undergo a change in its living conditions. Placing a bird outdoors is no problem if it has come from another birdkeeper's outdoor aviary, but it is a different matter if it has been kept indoors. This transition can be made quite safely from May and throughout the summer months, but it gets a bit risky after August, as then a bird that is not used to outdoor conditions no longer has the time to acclimatize before the winter sets in. The same applies to recently imported Aratingas. These will have just come out of the totally different weather conditions of a (sub-)tropical region. Therefore they must be given time to become accustomed to our climate before they go into the autumn and winter months. Even placing them outdoors in August is rather on the late side. In any case a close eye must be kept on any imported birds which are spending their first British winter in an outdoor aviary, as these will never be fully acclimatized. If you get the impression that they cannot yet cope with the cold and damp weather (ruffled feathers and half-closed eyes) take them indoors just to be on safe side. One advantage of Aratingas in this regard is that they virtually all spend the night in the nest box where it is always less cold and more sheltered.

V. Diet

We all wish to supply our Aratingas with the best possible diet, and nothing would be better than to first look at what they have on their menu in the wild. Unfortunately however, not much is known about this. It is possible to find out what sort of habitat they live in, but ornithologists and naturalists have very little to say about the Aratingas' feeding habits. The scanty information usually mentions various sorts of ripe and half-ripe seeds, with in addition many sorts of berries and fruit. They are also known to eat flower buds and heads, leaves and other vegetation, nuts and insects.

Though limited, this information is sufficient for our ends. They can easily be supplied with all these things in the aviary, and although they might not be exactly the same seeds, fruits and so on, their nutritional value will generally speaking be about the same. And even in their natural habitat certain species fill their stomachs with cultivated plants; they can sometimes even ruin the harvest.

Although Aratingas are able to survive for many years on nothing but seeds, it is of course not advisable to take this approach. The effect would certainly be noticeable in the breeding results; a varied diet produces many more offspring.

The necessary nutrients are proteins, fats, carbohydrates, vitamins, minerals and trace elements, and water. In the first instance these can be provided in the form of seeds (also germinated), fruit, berries, greenstuff (including a variety of herbs and herb seed), egg food, (willow) branches, calcium, grit (for grinding the food) and drinking water.

Generally speaking the seed mix which is available for the larger Australian parakeets is suitable for your Aratingas. However, you can also buy the seeds separately and mix them yourself. In order to keep the price down you will usually have to buy quite large quantities of each sort.

It goes without saying that the seeds should always be fresh. You can check this yourself by trying to germinate them. Old seeds germinate very badly and are of little or no nutritional value to your birds.

Parrots are very much partial to germinated seed; it is not only very nutritional, but it is also easy for young birds to digest. Seeds can be germinated by first placing them in plenty of water for 18 to 24 hours. Then, after they have soaked up enough water, the excess is poured away and the seeds are placed in a warm dry place. The first seeds sometimes begin to sprout after only one day, but the length of time depends mainly on the temperature. Always make sure that the seeds remain clean; the damp warm conditions necessary are also ideal for bacteria and mould. Rinsing the seeds regularly is therefore absolutely essential.

If you wish to germinate large quantities it can best be done by making lots of small holes in the bottoms of a number of buckets with, for example, a red-hot knitting needle, and then placing them inside each other. Make sure that the holes are smaller than the seeds, otherwise they will wash through. These are then placed in an untreated bucket in order to collect the surplus water. Using this through-flow system the seeds can be kept damp by occasionally pouring water into the top bucket. The seeds can dry out quite quickly, particularly when it is hot, and this adversely affects the germination process. If you put a portion of seed into soak, for example, every other day, then there will also be seeds ready for consumption every other day.

It is possible to use a standard seed mix for germination. However, these commonly small seeds often do not all germinate at the same rate, and the results are often disappointing. Therefore there is something to be said for using something else. I personally find a mixture of sunflower seeds, mung and other assorted beans, peas, dari, wheat and oats very successful. The birds eat this with relish. You can of course make up your own mixture.

Besides seeds, fruit and greenstuff are of incalculable importance to Aratingas, and may be allowed to make up 30% to 40% of their diet. Oranges, apples, sweet corn (preferably on the cob), carrots, beetroot (including the leaves), spinach and suchlike can form the basis.

In addition grapes, plums, pears, bananas, peaches, lettuce, endive, cabbage, curly kale, grass and various weeds (particularly chickweed) can be used. In fact almost everything is suitable, except of course for plants which are poisonous or have been sprayed with chemicals. If you possess a vegetable garden there is no end to what you can get out of it. Vegetables from other sources should be thoroughly washed.

Make sure that these ingredients are included in the diet, as they are extremely beneficial to the breeding results.

Egg food can be mentioned as the third component. Personally I always supply the seeds separately and then mix the egg food, fruit, greenstuff and germinated seeds together. With a diet such as this supplementary vitamins are unnecessary, at least as long as the birds eat everything properly.

The amounts which should be given of the various components depend upon the circumstances under which the birds are being kept. For example, birds which spend the winter in an unheated outdoor aviary need much more energy than those of the same species which are kept in a heated area indoors. This means that the composition of the diet should not be the same for both birds.

Every birdkeeper really has to work out what is best for his birds in their specific situation. There are a number of good books available on this subject which can provide clear and sound information for the interested aviculturist. However, several general basic principles can be mentioned here:
- the diet should consist of about 17% protein;
- the percentage of protein should be greater in the breeding period than outside it;
- when birds are in moult or laying they need extra calcium for the production of new feathers and eggshells;
- the smaller the birds' living quarters, the less fat their diet should contain;
- active birds can receive more seeds with a high fat content than quiet ones as they burn it up faster;
- birds need more energy in a cold period than in a hot one (energy is supplied by fats and carbohydrates);
- animal fats are necessary throughout the year under all circumstances.

It can only be to your own advantage if your breeding birds enjoy a varied diet. The goodness will not only be passed on to their young but they will also accept it much more readily later, as after fledging they will follow their parents example. What is more, a wide variety prevents the birds from receiving too much of one of the ingredients. It is worth bearing in mind that young birds are more willing to try new food than old ones. However, this is not generally such a problem, as Aratingas are inquisitive by nature and examine everything. If a bird's eating habits do remain one-sided it is a good idea to make a mixture which is not too recognizable. If, for example, it eats apple but not egg food, cut the apple into small pieces and stir it through this food. If a couple of pieces of apple are then placed on top, the bird will soon realise that there is also apple in the mixture.

In the above a few basic principles for your Aratingas' diet have been summarized. Of course this does not mean that they are hard and fast rules. Each birdkeeper tends to develop his own feeding system and is always in search of the one which proves most successful. As a rule once this success has been achieved (in the form of robust healthy young) it is advisable to stick to the same feeding system. However, it does of course make sense to remain open to the positive experiences of fellow aviculturists, and to occasionally be prepared to have a critical look at your own methods.

In order to offer some help in this matter, in the separate descriptions of each species in chapter IX I mention the diets which successful breeders have given their birds. I offer no opinions about the question of the suitability of these diets, but I think it is very much worthwhile passing on the information. You can absorb the knowledge, and you may possibly be able to use it to your advantage. After all, the time will never come when we are no longer able to learn anything from others.

VI. Breeding

In the wild Aratingas travel around in flocks for the greater part of the year. During the breeding season they split up into pairs. Although the moment at which this occurs is dependent on the climatic conditions and the food supply. In Central and South America these factors are less closely related to the seasons than is the case here in Europe. For example, the changes in temperature are less and the number of hours of daylight fewer.

It is a mistake to think that all Aratingas nest only in hollow tree trunks and branches. They do in fact also use other sites, such as termite hills (especially those in trees), rock crevices, hollows in cacti, and sometimes even in concealed holes in the ground.

It is striking that the members of this genus have little or no courtship behaviour. In fact generally the only activity apart from the actual mating is the ritual feeding. However, this can sometimes be very intensive and extend over a fairly long period. It regularly occurs that the hen suddenly disappears and it transpires that there are eggs; compared to the Australian parakeets it is much more difficult to spot when these species have reached this stage. For example, the cock never chases after the hen. If you pay careful attention, chewing the outside or inside of the nest-box can be an indication, and they may also spend more time the box during the day. Further, their behaviour towards their neighbours, and sometimes even towards their keeper, becomes much more aggressive.

Generally speaking, Aratingas accept partners fairly readily. In the great majority of cases two birds which have been placed together become companions, and are soon not only sitting side by side on the perches, but also preening each other. If this behaviour persists for a number of years without the production of young or eggs, it is advisable to split up the 'pair' as the chance of breeding success will only become slimmer. Of course birds which have recently been placed together need to be given some time to get used to each other (although this is virtually never a problem, and even love at first sight is not unheard of). A pair living in close harmony does not necessarily immediately produce offspring in all cases; all other conditions must also be optimal.
Newly-imported birds may require a number of years to adapt to their aviary environment, even though they are otherwise a well-matched pair. Younger birds are better able to adapt to such a situation than older ones. It goes without saying that things are much simpler if the birds have been reared in captivity. The birds must in any case feel safe and secure in their quarters. Keeping more than one pair of a species can have a favourable effect on the breeding results; they can encourage each other if housed within sight or shouting distance. It is better not to place them directly next to each other, as they may then start to squabble, and so lose their concentration on the matter of reproduction.

Generally speaking it can be expected that Aratingas must be two years old before they are sexually mature and able to breed. With a few species, such as the Queen of Bavaria's Conure, this may take even longer.
They have no specific courtship display, but a very strong pair bond does develop and the

cock feeds the hen a great deal. Mating takes place in the typically South American fashion: the cock remains sitting on the perch next to the hen, places one foot on her back and manoeuvres his belly into position against hers. It appears that some pairs also mate inside the nest-box. Birds sometimes mate purely in order to strengthen the pair bond.

The time that elapses between the first egg and the start of incubation differs for each pair. Hens occasionally start to sit immediately, and others only after laying the third egg. The majority start after two. The cock remains close by the nest-box, and he sometimes keeps the hen company a lot of the time. Cocks often spend the night on the nest with the hen. However, they never take part in the incubation; they leave that to the hens.

Although, as has already been mentioned above, Aratingas do not nest solely in hollow trunks and branches, in aviaries you can generally do with man-made or natural nest-boxes. In fact the birds accept these without any problems. Many pairs have the habit of chewing the inside of the box during the breeding season, thus lining the bottom with wood splinters, on which the eggs are then laid. The chewing usually ceases as soon as eggs have been laid. However, keep a close eye open while the birds are at it, as before you know it they can have chewed through the a wall of the nest-box, or worst of all, even through the bottom. This is something that I have had with, among others, my Jendayas. If you are not aware of the situation in time you run the risk of having eggs lying on the floor of the aviary, because the bottom has disappeared out of the box. This is of course a great shame, all the more when it can be avoided by a little attentiveness. If you have birds with this awkward habit it makes sense to take the precaution of covering the bottom of the box with a piece of fine wire mesh by attaching it to the sides, just in case.

If a pair is new it is a good idea to hang at least two different sorts of box in different places, so that they can choose. Once the choice has been made the other nest-box(es) can be removed. I need hardly mention that boxes should be securely attached.

Aratingas are not really fussy when it comes to nesting material. It is best to imitate the natural conditions by using rotten wood. This can virtually always be obtained somewhere. Coarse sawdust is also sometimes used (fine sawdust is too dusty, and can get into the chicks beaks, eyes and nostrils), and a thin grass sod can also be placed upside down inside the nest-box. Cover it with a layer of coarse sawdust and then make a hollow with your fist. Finally, some birdkeepers go into the woods and collect material containing old pine needles and/or rotting leaves from the ground. In the spring this will have become an excellent humus-like material.

Clutches usually consist of three or four eggs; sometimes increasing to up to seven for the smaller species. They are normally laid every other day; although a gap of three days is certainly not unusual. The incubation time is 23 to 25 days and depends somewhat on the weather conditions: the hotter it is the shorter the incubation period, and the colder the longer it takes.

Most pairs produce one clutch a year, although two is certainly not unheard of, particularly if a clutch proves infertile and the eggs are removed in time.

A fertile egg becomes darker the longer it has been incubated. After only four or five days a few veins begin to develop, which can be seen by holding the egg up to a light source. The darker the surroundings are, the better these can be seen.

As mentioned before, under normal circumstances fertile eggs hatch after 23 to 25 days. However, everyone discovers sooner or later that things do not always go according to plan. There are a number of causes for eggs failing to hatch, some of which cannot be helped. The eggs may become cold because the hen leaves the nest for some reason just before nightfall, and is then unable to find them again in the dark. If she stays away too long the embryos die. If the hen is too young, the breeding instinct may not yet be fully developed. Or one of the parents may be carrying a virus, which gets into the eggs and results in the death of the chicks. Or the shell of an egg may become cracked or broken, causing it to dry out. However, if you notice this in time it can be remedied by painting the spot with nail varnish. This seals the cracks thus halting the drying-out process and allowing the chick to continue developing. Put an egg which has had this treatment back in the nest only after the varnish has completely dried, otherwise there is a risk that the hen will start sitting again immediately and that her feathers will stick to it. An egg can also dry out if the shell is too thin and porous; this can normally be prevented by supplying the birds with sufficient calcium. Dead-in-shell eggs can also be the result of parents receiving such a poor diet that the chicks are not strong enough to break through the shell, or if they do manage it they remain much too weak and die later. Finally, there may be another reason for eggs drying out, for example exceptionally hot weather, but this is not a problem on the whole.

If in your opinion the eggs are taking too long to hatch there is a simple test to see if the chick is still alive: put the eggs into lukewarm water. If the chicks are still healthy you will see the eggs moving, but if they have died the eggs will remain motionless.

3. Dark clouds above the Ecuadorian rainforest, the home of the, among others, Weddell's Conure.

Do not become worried too soon if a chick has started tapping through the shell but then fails to hatch immediately. The time between the first crack and the moment that the chick actually frees itself from the shell can be as much as forty-eight hours. It should not be forgotten that this is a strenuous task for such a small creature, and allow it therefore to take its time. If you hold the egg up to your ear an excited cheeping can often be heard, and as long as that is the case there is nothing the matter. Therefore, do not take action too soon and bear in mind that there is a greater chance of causing damage than of preventing it. Experience has taught us that chicks which receive 'help' during hatching often fail to survive, so let nature take its course as far as possible.

Neither should you be worried if the chick does not have any food in its crop to begin with. A newly-hatched bird takes enough nutrition out of the egg to last it for at least twelve hours, so in the beginning it receives little or no food from its parents.

Chicks just out of the egg are pinkish in colour and are covered in off-white down. The eyes are closed and begin to open after about twelve days. When the birds are between ten and fourteen days old the dark feathers shafts become visible under the skin, and the beak and legs also turn a darker colour. After three or four weeks the first colours start to appear in the feathers, and they are almost fully fledged at an age of seven weeks, although some down is occasionally still visible here and there. It is around this time that they prepare to leave the nest-box. The exact moment is dependent upon particularly the weather and the diet.

They fly confidently almost immediately. They are even able to return to the nest-box for the night, and do so without fail. This represents a clear difference between them and the chicks of Australian parakeets, which are much more awkward after leaving the nest, are sometimes hardly able to fly, and never return to the nest-box.

Three weeks after leaving the nest immature Aratingas become independent. They may remain with their parents for some time; in the wild these birds frequently form family groups. However, in an aviary this can be risky if a second clutch is laid, as the entire family tends to spend the night together. If the nest-box contains eggs they may well be damaged.

The question of how sensible it is to inspect the nest-boxes will be answered differently by different birdkeepers. I do it fairly often, if possible every day. However, I never go so far as to chase the bird off the eggs. The important thing is to have a thorough knowledge of the habits and character of each separate breeding pair, and to adjust the frequency of the nest controls accordingly. It is not to be recommended if the birds are shy, and some caution is also called for if birds which have been recently acquired are breeding for the first time.

The great advantage of frequent nest control is that you can take action as soon as it is necessary. This can be in the case of, for example, a cracked egg, an egg which has rolled into a corner of the box, the hen not sitting properly, or the chicks not being properly fed. An infertile clutch can also be removed after between ten and fourteen days. After all, it is pointless to let a hen sit for a good three weeks for nothing.

It is a good idea to have an incubator in reserve for emergencies. The eggs can be temporarily placed in it if either the hen for some reason suddenly abandons her task, or lays too many eggs, or if for some other reason you are not happy with the situation. The temperature in the incubator should be between 99.4 and 100.5 degrees F.

If the problem is later solved the eggs can be replaced in the nest. In other cases there may be a bird available which has just laid eggs and would make a suitable foster mother. For this sort of temporary use, a simple and comparatively cheap machine can give excellent service.

Various crosses between species and sub-species of Aratingas have already appeared. This aspect of the hobby serves no purpose, and the only possible justification for it could be in order to answer the question of how closely related certain species are. However, there does not seem to be anybody who is prepared to tackle the question in a responsible manner. Whatever you do, avoid the use of such crosses as breeding stock, and make sure that none come onto the market.

In a number of cases recognizing crosses calls for a good deal of expertise, as some subspec species and sub-species are very similar in appearance. This is particularly true of the green Aratingas and the sub-species of the St Thomas Conures, the so-called Brown-throated Conures. As a guide I have included a table in the introduction to chapter IX (Description of species), in which the differences between the green-coloured species of Aratingas are described as clearly as possible, and further on, in the description of the St Thomas Conure, a summary of the various subspecies is given. By making use of this information it is possible to discover which species or subspecies a bird belongs to.

However, even with the summary this will not be easy, particularly where Brown-throated Conures (the sub-species of the St Thomas Conure) are concerned. This is because the differences are sometimes extremely slight, and also because you can never be completely certain that you are not dealing with a cross between two sub-species. A birdkeeper who has a single bird will search for a second, however, if he fails to find one he may make do with a bird which resembles his but which possibly is a member of a sub-species. Naturally, the young of such a pair are no longer recognizable as being pure-bred specimens of a sub-species.

VII. Diseases

Unfortunately sick birds are an unavoidable part of our hobby. At the same time they also form one of its most difficult aspects. After having gained a little experience everyone is able to distinguish a healthy bird from an ill one. But this is only the beginning, for then you have to find out what is wrong with it.
The creature is not able to explain what is wrong with it, so you have to rely on your own observations, or call in the help of a vet. They are often able to diagnose the problem by means of an examination, but in some cases it comes too late. If, for example, a bacterial culture has to be made the results will not be known for some days. The patient, however, cannot do without medicine in the meantime, for in many cases illnesses of aviary birds run their courses extremely quickly. A bird which is ill one day may be dead the next unless the correct medicine is prescribed.

It is regrettable that the hobby is still in its infancy with regard to this aspect. This is due in the first place to the fact that there are only a handful of vets who have a thorough knowledge of parrotlikes, and who are clearly interested and/or specialize in them, and in the second place because the great majority of dead parrots are not made available for a post-mortem. This would make it possible to collect a wealth of information.
Moreover, for one reason or another probably too few people take their sick birds to the vet. This, of course, gives him no reason to start concentrating on this catagory of patient. This results in a vicious circle which cannot be broken.
Luckily there are some new developments but things are changing only very slowly. However, the process can be speeded up by regularly visiting a vet who in your opinion is good, and also by encouraging your fellow hobbyists to do likewise. Everyone will undoubtably profit from this in the long run, all the more as you will then be able to share your experiences with others.

As the subject in hand is really only for specialists it falls outside the scope of this book, and I will not say much about it. The aim of the book is, after all, to show you the ropes by supplying as much information as possible about the species within the genus *Aratinga*. One of the ropes is disease, but this problem is common to all birdkeepers and the same diseases strike all species. It is therefore advisable to purchase one of the number of books on the market which are specially dedicated to the subject.

It is certainly not an easy subject, but it is part and parcel of the hobby, and it is therefore a good idea to have at least some knowledge of it. If you do not wish to go into it too deeply, you can of course always go to a vet or a fellow birdkeeper in the majority of cases. However, there will undoubtably be the odd occasion when for one reason or another these are not available, and that you will have to give a particular bird treatment immediately. It is at times like these that it is to your advantage if you have some idea what goes on within the bird's body, and about possible treatment. It can be seen as a sort of first aid.

With this in mind there follows a short summary of what every birdkeeper should have at their disposal, and what they should be able to do; for further information I refer you to the more specialized literature.

Measures and requirements

1. Hygiene.
Make sure that your aviary is properly clean to prevent your birds from being bothered by unwanted visitors such as red mite, and feather mite. It is advisable always to have a suitable remedy at hand, and also a disinfectant such as Halamid.

2. A net and gloves.
It is better to handle parrots without gloves on as you can then feel if you are holding them properly; however, not everone feels confident about doing this.

3. A hospital cage or a wire cage with an infra-red lamp.
One of the first things that a sick bird should receive is warmth, and the temperature should reach between thirty and thirty-five degrees Celcius. Hospital cages often have the drawback of being closed, which makes them dry and dusty and the temperature within them even throughout. A simple wire cage with a lamp attached on the outside does not have this problem: there is more fresh air and the patient can move closer to or away from the heat source (the lamp) in order to find the most suitable temperature. An advantage of an infra-red lamp is that it is not either constantly light in the cage, or, if a thermostat is used, that the light is not continuously turning on and off.

4. A haemostatic (blood-clotting) agent.
For example iron chloride, for the treatment of open wounds.

5. A pipette or crop needle to administer medicine.
The most reliable method is to use a crop needle, as you are then certain that the bird has taken all of the medicine. These are available from a vet.

6. A widely applicable antibiotic such as Vibramycin.
Antibiotics come in various forms, and they are not all effective against the same bacteria. If you use a broad-spectrum antibiotic (effective against a large number of bacteria), the chance of a cure is greater. However, if you do not know exactly what is wrong the use of such an antibiotic is no more than guesswork, and then it is better to have the bird examined in order to find out what precisely the problem is. Only then can the correct medicine be prescribed and administered.

7. A dewormer.
One of the first things regarding bird health that every beginner discovers is that parrots are susceptible to all sorts of worms which build up in the intestines. At one time many birds died of these infestations.
Nowadays they do not need to form any problem as their treatment has become an integral part of keeping parakeets. A good dewormer which is available from your vet is Fenbendazol, a white liquid of which the prescribed number of drops can be administered to

each bird (in the beak with a pipette or into the stomach with a crop needle).

You can also make up a very cheap medicine by buying the powder Panacur from your vet. This also contains the aforementioned Fenbendazol as active ingredient. You mix 5g of this with red grenadine or diluted orange squash until you have 20cc of liquid (1g of Panacur is a slightly rounded teaspoon). This 20cc is sufficient for 400 drops; 1cc therefore contains 20 drops. You need two drops of this solution for each 40g of bodyweight; in other words a bird weighing 40g receives two drops, a bird weighing 80g four drops, a bird weighing 120g six drops, and so on. In chapter IX the weight of each species is given, so that you can easily work out how much each bird needs.

The only drawback with this mixture is that it can only be kept for a few months in the refrigerator, as after that the fermentation process begins. However, this is no real problem; you just make up the necessary amount and if there is any left over you can throw it away. This will be a waste of only a few pence. In powder form Panacur can be kept for at least two years.

If you dislike having to catch your birds for deworming there is also a simpler method, although it is less reliable. You put a small amount of Panacur powder onto something which the parrot is very partial to, for example a piece of apple, orange or millet which you have made damp. As long as one bird eats the whole dose you can be reasonably certain that the treatment will be successful. The difficulty is that two birds are often kept in one aviary, and then it is often difficult to tell if they have both eaten the treated food or that one of the two has had a double dose. This problem can be solved by shutting up one bird at a time in the night shelter and serving it the treat. The cure should be repeated two or three weeks later, as worms eggs are not always completely killed off, and any which may have been developing at the time of the first cure will then have come out.

VIII. Administration

It is very important to keep your birds' particulars up to date. Even if you only have a few pairs it is a good thing to note down any information, as even then it is difficult to remember everything and to know exactly which information belongs to which bird.

This can be done in various ways. The first possibility is to buy a new diary each year and to write down by the relevant date the thing you wish to remember. The drawback with this is that you never have all the particulars together but always have to search for them. Therefore, it is better to use a system whereby you note the necessary information for each pair separately. I personally have been using the following table for years, two of which can be typed on one A4 sheet of paper. This can then be copied as often as is necessary, and a new table should be used each year. The particulars for two clutches can be filled in on each table.

By putting the tables of two pairs next to each other you have an excellent summary of everything you wish to know. Particulars such as the size of the aviary, the type and size of the nest-box, the size of the entrance hole, the type of nest material, etc. can of course be added to the table, or others scrapped from it, according to individual needs.

SPECIES: _____ Number: _____ Aviary No: _____
_____ Ring size: _____ mm Year: _____
COCK Ring no: _____ Year of birth: _____ Origin: _____
HEN Ring no: _____ Year of birth: _____ Origin: _____

Dates eggs laid _____ Special comments:
Number of eggs _____
Number fertile _____
Dates eggs hatched _____
Number hatched _____
Ringing dates _____
Ring numbers _____
Dates chicks left box _____

Dates eggs laid _____
Number of eggs _____
Number fertile _____
Dates eggs hatched _____
Number hatched _____
Ringing dates _____
Ring numbers _____
Dates chicks left box _____

The table speaks for itself. Under "Special comments" you can note anything which you find interesting but which does not come under any other heading in the table. From the other particulars you can work out what the incubation time was and the number of days which passed before the chicks left the nest-box.

When most birdkeepers begin, they think that they will be able to remember everything. However, time and again it has been proved that the longer you try to do this the more trouble you will have to remember everything. The situation becomes hopelessly difficult if you have more than one pair of a species, as the particulars are then easily confused. If you wish to carry out your hobby seriously, it is essential to be able to refer to reliable information on each separate pair: how many eggs there were in the clutches, how long there was between the laying of each egg, how many were fertile, what the incubation time was, how many hatched, how the chicks developed, how the parents fed the chicks, and when the chicks left the nest. In this fashion a wealth of information can be built up over the years, which now and then can be looked through with pleasure. Moreover, you should not forget that it can be of great help to your fellow hobbyists.

IX. Species descriptions

In this chapter each of the nineteen different species within this genus is dealt with separately. A further thirty-five subspecies belonging to these nineteen species, one of which has died out, makes a total of fifty-three different birds which are of potential interest to birdkeepers. All the birds are discussed, although some are gone into more deeply than others, depending on how much is known about the species in question. Some are known to everybody, whereas most birdkeepers have never even seen others.
So far as I been able to ascertain, members of at least the following species are to be found in aviaries in Europe: Sharp-tailed, Queen of Bavaria's, Finsch's, Wagler's, Mitred, Red-Masked, White-eyed, Golden-headed, Jendaya, Sun, Weddell's, Jamaican, Petz's, St Thomas, Cactus, Golden-crowned and Cuban. That is no less than seventeen of the nineteen. The missing two are the Mexican Green and the Hispaniolan.

Within the genus several separate groups can be distinguished:

- a. there are three species which form a somewhat independent group, and which are probably not very closely related to any of the other species. These are the Sharp-tailed, the Queen of Bavaria's and the Weddell's Conure.
  Although the posture and the colouring of the Sharp-tailed is somewhat similar to that of the green-red group mentioned in d, it also closely resembles several of the dwarf macaws, such as the *Ara nobilis*, which suddenly appeared on the Dutch market in 1988.
  The behaviour, colouring and the shape of its beak, tail and body give the Queen of Bavaria's Conure a unique position within the genus. It is sometimes even doubted whether its inclusion into the genus *Aratinga* is correct.
  Finally, the Weddell's Conure shows some similarity to the group mentioned in c. But particularly the head and the striking white eye ring lend it a special appearance;

- b. Golden-headed, Jendaya and Sun Conure. The relationship between these species will be obvious to every birdkeeper; some even claim that the three together form a single species. Their various distribution ranges do not overlap at any point. This ensures that they do not have to compete with each other in matters of reproduction and feeding;

- c. Jamaican, Petz's, St Thomas, Cactus and Golden-crowned Conure. The distribution ranges of these five species fit together like pieces in a jigsaw puzzle, so they also do not form competition for each other. For the rest, the similarities in their behaviour show that they stand very close to each in the biological order;

- d. Mexican Green, Finsch's, Wagler's, Mitred, Red-masked, White-eyed, Hispaniolan and Cuban Conure. In respect to distribution range and behaviour, the same applies here as for the group mentioned in c. These species also have in common that they are principally green and have varying amounts of red on their heads and the front edge (alula) of the wing. Most birdkeepers are not able to identify them all accurately. For this reason I have pointed out the most important differences between the species in the following table. When you have used it to find out which species you are dealing with, you can turn to the relevant species description in order to find out which subspecies it may be.

species	length (cm)	front	crown	bend of wing	thighs	lesser under-wing coverts	greater under-wing coverts
A. h. holochlora	32	green	green	green	green	yellowish-green	yellowish-green
A. finschi	28	red	green	red	red	red	yellow
A. w. wagleri	36	red	red	green	green	green	olive yellow
A. m. mitrata	38	red	green	green	green	olive green	olive green
A. erythrogenys	33	red	red	red	red	red	olive yellow
A. l. leucopthalmus	32	green	green	red	red	red	golden yellow
A. c. chloroptera	32	green	green	red	green	green	red
A. euops	26	green	green	red	green	red	olive yellow

Textual divisions

In the following species descriptions the text is divided up into sections under the following headings: sub-species, origin of name, parents and young, sizes and weights, habitat and habits, diet, nesting sites, breeding process, general remarks, and mutations. Here is a summary of what information is included in each section.

Sub-species

Not all species have any sub-species. If that is the case, the scientific name only consists of two words, for example *Aratinga guarouba*. The first word indicates the genus, the second the species. If there are sub-species the name consists of three words, for example *Aratinga pertinax surinama*. The first two indicate respectively the genus and species, the third the sub-species. If the last two words are the same, for example *Aratinga pertinax pertinax*, this is called the nominate form.
If a species has sub-species, these are named, and the differences between them are mentioned. They are all mentioned separately and the scientific and the English names, if any, are given.

Origin of Name

The meaning of the scientific name is explained in this section. Besides the English name, the Dutch and German names are also given.

Parents and Young

As a photograph of each species is included, it is not necessary to give a full description of the plumage. For this reason, only additional comments are made regarding the appearance of the parents and immature birds, and any possible difference between the sexes.

Sizes and Weights

The length, weight, and ring size are mentioned. With regard to length, that of the nominate form is always given first. If under 'weight' the figures given are marked 'approximate', it means that they are approximations based on information from other aviculturists. Other weights given, sometimes one and sometimes more, are those of birds which I have weighed myself. Ring size needs no explanation.

Habitat and Habits

This information is given in combination with a map showing the distribution range of each species, and concerns the natural habitat in which the species is found and the behaviour that it shows in the wild. This knowledge can be of advantage to every birdkeeper with regard to creating a suitable aviary environment.

Diet

A distinction is made between the wild and the aviary. The food which is eaten in the natural environment is mentioned first. There then follow one or more menus which keepers can serve their birds. These have all been taken from reports on rearing the species concerned. I did of course come across some which I considered to be unsuitable, and these have been left out. But even then, it is not so that all the menus given are necessarily perfect and that you should follow them to the letter. They are rather meant to give an idea about how fellow aviculturists care for their birds, and to provide potentially useful tips.

Nesting Sites

Information is given about the nesting site which the species in question uses, both in the wild and in the aviary. The particulars given about the aviary are mainly taken from successful breeding results.

Breeding Process

It is important to know how many eggs are laid, the incubation time, when the young leave the nest-box, etc.

General Remarks

These include information which does not fall under one of the above headings. This usually includes how long the species has been known as an aviary bird, when and in which country it was first reared, etc. The breeding results in the Netherlands are given as far as possible. Further, any other important additional information is included when necessary.

Mutations

All known mutations are mentioned. However, with regard to conures, birdkeeping is still in its infancy as far as this aspect is concerned.

4. Sharp-tailed Conures are very similar in appearance to the Dwarf Macaws.

SHARP-TAILED CONURE - *Aratinga acuticaudata acuticaudata*

Sub-species

1. *Aratinga acuticaudata acuticaudata* - Sharp-tailed Conure
2. *Aratinga acuticaudata haemorrhous*   - Blue-crowned Conure
3. *Aratinga acuticaudata neoxena*   - Margarita Conure
4. *Aratinga acuticaudata neumanni*   - Bolivia Conure

The differences between the sub-species are to be found mainly in the amount of blue on the head, and in the colour of the beak. The upper mandible of the nomimate form (no. 1) is the colour of bone with a dark point, and the lower mandible is dark. Moreover, this is the only one with a completely blue head: forehead, crown, lores, cheeks, and ear coverts.

Number 2 has only a blue forehead, and it is much paler than that of number 1. The green is a little brighter, and both the upper and lower mandibles are bone coloured.

The forehead and crown of number three are a dull blue, the upper mandible is bone coloured and the lower is greyish-black.

Number 4 has a true blue forehead, crown and nape, and the green is brighter; the lower mandible is most likely dark.

Particularly number 1, and to a lesser degree number ? (zie bld 45), are known to aviculturists; number 2 is no longer found here because of the export ban imposed by Colombia, Venezuela and Brazil.

Origin of Name

Aratinga: shining or glittering parakeet.
Acuticaudata: sharp-tailed.
Haemorrhous: flowing blood.
Neoxena: new stranger, guest.
Neumanni: Oscar Neumann (1867-1946) from Berlin undertook several journeys to Africa to collect animals, and developed into a world-famous zoologist.
Dutch: blauwkoparatinga
German: Blaukopfsittich, Spitzschwanzsittich.

Parents and Young

The Sharp-tailed Conure is mainly green. The head has varying amounts of blue, and the underside of the tail is reddish-brown. It has a striking white eye ring. The upper mandible of all the sub-species is bone-coloured; the lower mandible is either bone-coloured or greyish-black. The colour of the plumage and the eye ring is paler in

immature birds, and the lower mandible is bone-coloured in all species. In all other respects they resemble their parents.

Silva suspects that they are only sexually mature after three years. However, it seems more likely to me that they are able to reproduce after only two years.

Sizes and Weights

Length: Sharp-tailed 37cm (14.5in), Blue-crowned 35cm (14in), Margarita 32cm (12.5in) and Bolivia 37cm (14.5in).
Weight: approximately 190g.
Ring size: 7mm.

Habitat and Habits

Sharp-tailed Conures are common in the wild. They are birds of the comparatively dry regions which are able to adapt fairly easily to changing circumstances, but which avoid the humid woodlands. Outside the breeding season they travel around in flocks, and prefer to keep to open woods and areas of brushwood, although they will also go into open savanna and even semi-desert. They are principally birds of the lowlands.

Little is known about the three sub-species. The Margaritan inhabits extensive savanna and the fringes of woods, and the Bolivian sub-species is a mountain bird and lives in the Andes at an altitude of between 1500 and 2700 metres.

In the breeding season pairs separate themselves from the rest. Their flight distance (?) is small, which means to say that in the wild they can be approached quite closely. They also become reasonably tame when kept in an aviary. It appears that young birds which are kept separately are even able to learn to speak a few words.

In flight these conures look very similar to the smaller macaws. It can be seen that confusion sometimes occurs from the fact that in Brazil and Paraguay they are regularly called 'maracana', a name which is also given to the dwarf macaws. In Argentina it is normally called 'loro de los palos' (parakeet of the branch), after its habit of often sitting in the topmost branches of dead trees. They are often found there in large groups together with Mitred Conures (*A. m. mitrata*).

Diet

Wild: they can sometimes cause damage in grain fields, but on the other hand they make themselves useful by eating seeds of weeds. In addition they are fond of the fruits of, among others, cacti and mango trees, berries and nuts.

Aviary:
1. besides a seed mix, give daily apple, orange, spinach, beetroot, peas, carrots or other fruit and greenstuff which is in season. Germinated seeds are also a good source of nourishment, and further pigeon feed can be added to the list;
2. various seeds and peas, always soaked; mixed daily with fruit and vegetables, among others apple, orange, pear, plum, melon, sweet corn, beans, cucumber and celery.

Nesting Sites

The nests are made in holes in trees. One has even been found at a height of seven metres. Other than this little is known.
In aviaries they have been bred both in standard Cockatiel nest-boxes with a thick layer of wood shavings on the floor, and in boxes measuring 30x30x45cm (12x12x18in). However, it is better to use a box 50 to 60cm (19.5 to 23.5in) in height, and 15 to 20cm (6 to 8in) square. The entrance hole should have a diameter of 8cm (3in).

Breeding Process

The American aviculturist, Robbie Harris, keeps Sharp-tailed Conures in a cage measuring 76x46x46cm (30x18x18in) with a Cockatiel nest-box hung on the outside of it. Her pair breeds once a year. The first time three eggs were laid, on 7th, 9th and 11th June; all fertile. Only the hen incubated, and the cock only entered the nest-box when it was dark. The chicks hatched on 30th June and 2nd and 5th July. The incubation time was therefore 23-24 days. The chicks have light down; after a week they weigh about 45g. The eyes open on the eleventh day, and at an age of three weeks feathers appear on the tail and wings. The beak is then beginning to turn from flesh-coloured to grey.
Thomas Arndt's birds in Germany always breed in May and lay three eggs. The chicks are fully developed within 50 days.
The birds of Tony Silva gets on average three more, which according to him are incubated by the hen for 26 days. Newly-hatched chicks have white down. The first feather follicles appear after six days and the claws are turning grey. In his case they leave the nest-box after about 57 days.
I myself had chicks only two years after purchasing birds from the first shipment of these conures (*A. a. acuticaudata*) into the Netherlands in 1977, something which a fellow Dutchman Ruhof also achieved. He had even had two, albeit infertile, eggs in 1978. The following year things went better, for both his pairs laid three eggs in May and June in natural boxes 60cm (23.5in) high and 20cm (8in) in diameter. All the eggs hatched. However, the last one of each clutch to hatch was dead within a week.
In my case the four birds which I had purchased were placed in a spacious aviary measuring 3x7m (10x23ft), with an adjoining night shelter which was divided into three separate compartments 1m 80cm (6ft) deep and 1m (3ft 3in) wide. The first of three eggs was laid on 9th July, in a nest-box 60cm (23.5in) high with a floor surface of 20x20cm (8x8in). All three were fertile, and they measured respectively 26.0x33.1mm, 26.0x32.1 mm and 25.5x32.6mm. After she had laid the first egg the hen hardly ever left the box; she only came out every now and then to be fed by the cock. The first chick hatched on 2nd August, the second on 5th, and the third on the 7th August. This suggests that the hen had started incubating as soon as the first egg had appeared, and that the average incubation time was 24 days. I only ringed the first chick with a 6mm ring (which later proved to be too small) on the seventeenth day. My youngest chick also died; the other two left the box on 1st October; the sixtieth and the fifty-seventh day. It was only some years later, in 1983, that the Dutch Birdkeepers' Association (Nederlandse Bond van Vogelliefhebbers) awarded a certificate for an exceptional breeding success. The Dutch Parrot Society already had its first Dutch-bred bird at the 1979 show.
So far there has only been one clutch a year in all cases.

General Remarks

Specimens of this species (sub-species *A. a. haemorrhous*) were first brought to London in 1864. Some time later these birds had young. After that it was not until 1971 that the 'first' breeding success was reported, once again in England (number 1). The first success with number 2 was achieved in 1950 in California. As far as it is known the first members of this genus (number 1) reached the Netherlands in 1977, and from one year to the next they were suddenly freely available. Nowadays you hardly ever see them. Sharp-tailed Conures can sometimes be rather noisy, although generally speaking they are no match for, for example, Jendaya and Sun Conures. If they have eggs or young they are often more or less completely silent. My own birds were generally always fairly quiet.

Mutations

There are two splendid lutino hens in the collection of Nelson Kawall, a leading Brazilian aviculturist in San Paulo. The heads of these birds have beautiful apricot-coloured caps; a colour which is identical to that of the White-bellied Caique.
An aviculturist in Germany possesses a cream-coloured bird.

5. The splendid Queen of Bavaria's Conure is the most endangered *Aratinga*.

QUEEN OR BAVARIA'S CONURE - *Aratinga guarouba*

Sub-species

None.

Origin of Name

Aratinga: shining parakeet.
Guarouba: after the Indian word 'Guarùba' or 'Guarajùba', meaning 'yellow bird'.
English: also Golden Conure.
Dutch: goudparkiet.
German: Goldsittich.

Parents and Young

These birds have a simple, yet breathtaking and fascinating plumage of golden yellow and green. In this respect there is no difference between the cock and hen. Nor does behaviour give any guarantee of a bird's sex. The males may be slightly more heavily built, particularly with regard to the size of the head and bill, but even this is not true in all cases. Both sexes have relatively large bills.
Immature birds can be clearly distinguished from their parents: the yellow is duller and the plumage is dotted with green feathers, particularly on the head and to a lesser extent on the mantle, wings and breast. They develop their full adult plumage at about eighteen months old.

Sizes and Weights

Length: 35cm (14in).
Weight: approximately 250g.
Ring Size: 10mm.

Habitat and Habits

This conure comes from a northeastern area of Brazil where a great deal of jungle has already been cleared in order to build large motorways and their accompanying network of link roads. For two motorways, no less than 1,775,000 hectares (4,386,000 acres) of forest has been destroyed! The ecology of all the land within 50 kilometres (30 miles) of these motorways undergoes a change and becomes unsuitable for these parrotlikes. The

Queen of Bavaria's Conure is an inhabitant of tropical rainforest and is unable to adapt to a different environment; as a result, this species is seriously endangered, a fact which can be seen partly in the size of the flocks. These would normally contain between six and thirty birds, but more and more groups of two or three birds are being seen.

Other threats are, the building of dams and the flooding of large areas of the jungle for the production of electricity and the shooting of birds by farmers who want them out of their corn fields.

The adult bird's threat display involves a wide range of expressive movements: bill-tapping, beating of the wings, spreading the tail, swaying the head from side to side, putting the head under the wing, walking back and forth along a branch, etc.

Diet

Wild: fruits, berries, seeds and nuts, which are generally sought for in the canopy of trees.

Aviary:
1. a standard seed mix (consisting mainly of sunflower seeds, pine nuts, peanuts and a little kardi, oats, wheat and maize) with particularly in the breeding season lots of fruit (apples, oranges and bananas), a yeast rich in protein, vitamins and minerals, cottage cheese with wheat-germ oil and germinated seeds;
2. a seed mix, sunflower seeds (soaked and germinated), apples (with vitamin and a mineral preparation), brown bread with milk and honey and berries. They eat little in the way of greenfood.

Nesting Sites

Jim Hayward breeds these birds in plywood boxes 60cm (23.5in) tall and 23cm (9in) square, with a layer of rotten wood and sand on the bottom.

However, generally speaking the size and shape of the nesting site is not of great importance. Queen Conures sleep in their nest-box every night and for this reason they accept them readily. As they are sometimes very active chewers, it is better if boxes are thick-walled. This gives the added advantage that they are better insulated.

Breeding Process

In Western Europe these birds usually come into breeding condition in May and June, although, on odd occasions, eggs have been laid as early as February.

They mate in the typically South American fashion, which involves the cock placing one foot on the hen's back, while clinging firmly onto the perch with the other. When the pair are ready to breed they often become quite aggressive, and can even attack their keeper.

The birds belonging to Hayward usually produce four eggs, sometimes six. The first chick hatches 31 days after the first egg has been laid. The chicks have a light covering of white down and leave the box after nine or ten weeks.

The Swiss aviculturist, Etterlin, has measured a number of eggs: 24-25mm x 34-36mm. In his case the first egg, which was laid on 14 May, hatched on 10th June. On average the

hatching time will be about 25 days. The chicks left the box after ten weeks.
Parents and young form a close family group. In the aviary they can be kept together for up to three years without any problem.

General Remarks

The Queen of Bavaria's Conure was first described as early as 1648! In 1871 it was on show in London Zoo. The first breeding success was in Sri Lanka in 1939. In the Netherlands a Breeding Award was given for rearing this species in 1985.
On 3rd March 1973 it was placed in Appendix I of the Washington Convention.
Queen of Bavaria's Conures have two bad habits: one is their powerful voice, which is capable of developing an ear-splitting scream, and the other is their tendency to pluck their feathers. The cause of the latter is not known; some birds possibly receive too few stimuli and have too little distraction in the limited space of an aviary, which causes them to become bored and leads to 'preening' in this exaggerated fashion. It is probably the result of a behavioural disturbance. It really is a great problem, and in the worst cases they end up almost entirely naked. I have seen these birds in various aviaries, and in almost every case there was some sort of problem with the plumage.
They are fairly hardy and can put up with low temperatures, as long as they are able to spend the night in their nest-boxes. Unlike most Aratingas, they become very tame and intimate. Even birds caught in the wild become very calm after a time, without any particular attention being paid to taming them.

Mutations

None known.

MEXICAN GREEN CONURE - *Aratinga holochlora holochlora*

Sub-species

1. *Aratinga holochora holochlora* - Mexican Green Conure
2. *Aratinga holochora brevipes* - Socorro Green Conure
3. *Aratinga holochlora brewsteri* - Brewster's Green Conure
4. *Aratinga holochlora strenua* - Nicaraguan Green Conure
5. *Aratinga holochlora rubritorquis* - Red-throated Conure

This species is found in part of Mexico and Central America.

Number 1 is completely green, the belly tending to yellowish-green. Sometimes there are the odd red feathers on the head and neck.

Number 2 is very similar, but the belly is darker green and there are never any red feathers. According to Arndt the tenth feather in the wing is shorter than the seventh; this is the other way round in the other sub-species.

Number 3 is also almost the same as number 1, but has a darker green belly and the crown has a faint blueish tint.

The Nicaraguan (number 4) is also similar in appearance to the Mexican Green (1), but is larger with a heavier bill and sturdier legs.

The Red-throated Conure (number 5) is actually the only one which clearly distinguishes itself from the other sub-species by its red throat; this can vary in shape and sometimes extends into the neck.

Origin of Name

Aratinga: shining parakeet.
Holochlora: completely green.
Brevipes: small leg.
Brewsteri: William Brewster (1851-1919) was a leading American ornithologist.
Strenua: lively, energetic.
Rubritorquis: red necklace.
Socorro: island off the west coast of Mexico.
Dutch: groene aratinga.
German: Grünsittich.

Parents and Young

Number 1 is almost entirely green, merging into a yellow-green on the belly. Some birds have a few red feathers on the head and neck. This appears to occur more often with cocks than with hens. The bend of wing is green. The naked eye ring is flesh-coloured, the iris is orangeish-red.

6. *The Nicaraguan Green Conure only occurs in Central America.*

Immature birds are similar to their parents, but still have a brown iris. Red-throated Conure chicks emerge from the box entirely green; there are no red feathers on the throat at fledging.

Sizes and Weights

Length: Mexican Green 32cm (12.5in), Socorro 34cm (13.5in), Brewster's 32cm (12.5in), Nicaraguan 34cm (13.5in), Red-throated 30cm (12in).
Weight : Nicaraguan 140g, Red-throated 120g.
Ring size : 6mm (leg size of Red-throated 4.8mm).

Habitat and Habits

This species has the most northerly distribution range of all the Aratingas, that being Central America from Mexico to Nicaragua. All the sub-species avoid damp and low-lying rainforest, and prefer deciduous forests, shrubland, open areas and forest fringes. Outside the breeding season they travel about in groups, although the pair bond is clearly maintained within the group. The size of the groups depends on the availability of food, but usually numbers at least 20 to 30 birds.
The Mexican Green Conure lives in pine forests, on the lower hills and in the tropical plains towards the Pacific, and is therefore not really a forest dweller. The Socorro sub-species is mainly found in the woodlands growing at a higher altitude on this island. The Brewster's is a mountain bird which inhabits land between 1250 and 2000 metres. The Nicaraguan form can be found in the highlands and the dry interior of Guatemala and El Salvador, on the mountainsides near the coast up to an altitude of 1400 metres, and also in pine woods, more open wooded areas and agricultural land, where it can cause some damage. The Red-throated Conure lives almost entirely in high-altitude pine forests up to about 2600 metres, and is therefore able to withstand our low temperatures fairly well.
The Mexican Green is reasonably common and the Red-throated is common; little is known about the other sub-species in this respect.

Diet

Wild : the menu consists mainly of berries, seeds, nuts, greenery and fruits, the greater part of which the birds find in the canopies of trees and in bushes.
Aviary:
1. a seed mix for large parakeets, and in addition germinated seed, egg food, greenfood, fruit and berries.
2. a mixture of large seeds, pine nuts, and daily fresh fruit and greenfood (lots of tree leaves), also dandelion leaves, blanched chicory and carrot.

Nesting Sites

In the wild this species breeds in large holes in trees; the Nicaraguan Conure is also

known to dig holes for itself in the nests of tree termites, to lay and hatch its eggs.
The box de Ruygt (Holland) used (see below) was 50cm (19.5in) tall, measured 19x24cm (7.7x9.5in) inside and had an entrance hole 7cm (3in) in diamater. The walls were no less than 6cm (2.5in) thick.

Breeding Process

A report by de Ruygt about breeding the Red-throated Conure in 1986 concerns a clutch of four eggs, which were incubated for 24-25 days before they hatched. The chicks eyes opened on the twelfth day, and they were then ringed. The first left the box on the 54th day. Three young were reared, which were sexed at an age of 36 weeks: two cocks and a hen. The parents had spent six years in the aviary before this success was achieved.

General Remarks

The Mexican Green was first bred in the US in 1934, but so far probably not in Europe. The same is true of the other sub-species, apart from the Red-throated. As far as is known, the German aviculturist, Arndt, was the first to breed this species in 1976, and the Netherlands followed in 1979. This Aratinga is almost unknown as an aviary bird; the few pairs of Red-throateds being the exception. This lack of availability will not cause most birdkeepers concern because of its almost entirely green appearance little interest will probably be shown in this bird.

Mutations

None known.

7. Red-throated Conures.

8. *The Finsch's Conure can be identified by its yellow under wing coverts.*

FINSCH'S CONURE - *Aratinga finschi*

Sub-species

None.

Origin of Name

Aratinga: shining parakeet.
Finschi: after Dr Friedrich Hermann Otto Finsch (1839-1917), a German ornithologist who wrote, among other books, the standard work "Die Papageien", published in 1867.
Dutch: Finscharatinga.
German: Rotstirnsittich.

Parents and Young

The forehead is red, and in some birds can extend as far as the eyes; however, this is not often the case and usually there is a green stripe above the eye. There is sometimes a red band on the chin, but only the odd red feather is found. The greater under wing coverts are yellow, the lesser coverts are red. The bend of wing is red, and is clearly visible when the bird is sitting quietly on its perch. The iris is orange.
Young birds are similar to their parents but have very little red on the forehead and bend of wing.

Sizes and Weights

Length: 28cm (11in).
Weight: 155 and 180g.
Ring size: 6mm (leg size 5mm).

Habitat and Habits

This Aratinga inhabits Central America from southern Nicaragua to western Panama. It is normally found in lightly-wooded country, open areas, agricultural land, and even in villages and towns. It lives principally in the lowlands and low in the hills of these tropical and subtropical regions. Within its limited distribution range it is a common species, which is possibly on the increase due to the cultivation of the land. It rarely enters woods.
Outside the breeding season they travel about in groups of between ten and a hundred

birds. Within the group the various breeding pairs keep close contact. At their roosting sites, where up to 500 birds can sometimes be found together, they of course produce a deafening noise in the battle for the best places. In the aviary they are fairly quiet birds, and what is more they are not very destructive.

Diet

Wild: for the main part fruit, berries, nuts and various seeds. They also make grateful use of crops grown by farmers.
Aviary:
1. there is little information to be found about diets specifically for this species which goes further than those generally used for aviary birds. If you have these birds you can give them the same as the other green Aratingas;
2. sunflower seeds, mung beans and oats, also in germinated form, soaked pigeon feed, and a mixture of mung beans and pigeon feed which has been boiled for two minutes. A good egg mix can also be added to the last mentioned. Also boiled maize, pieces of fruit, berries and greenfood, and finally a seed mix of white sunflower seeds, kardi and millet. A multi-vitamin and mineral preparation is sprinkled over the food, along with calcium from mussel shells.

Nesting Sites

In the Walsrode Bird Park in Germany they have bred these birds in a cavity 60cm (23.5in) deep and 20cm (8in) in diameter in a natural log. The entrance hole was 30cm (12in) above the floor and was 9cm (3.5in) in diameter. The birds hardly chewed the wood at all. They would probably also use a nest-box between 40 and 50cm (16 and 19.5in) tall with a floor area of 20x20cm (8x8in). A 9cm entrance hole seems a bit on the large size; 2cm less should also be sufficient.

Breeding Process

Little is known about the breeding process in the wild. The only breeding success reported involving birds in captivity is that at the Walsrode Bird Park. Although chicks hatched in 1982 they were not fed and failed to survive, so the first real results were in 1988. Between 2nd and 9th June four eggs were laid, and incubation probably started after the second egg. The cock spent a great deal of time in the proximity of the nesting site, and joined the hen on the nest at night. The first two chicks hatched after 23 days, and a third followed a day later. The fourth egg was not fertilized. The chicks left the nest after 56, 56, and 55 days respectively, and were rather shy to start with. They had no red in their plumage at first. Their foreheads were fairly dark.

General Remarks

This species exhibits several similarities to the Mexican Green Conure, and according to some experts it is best to regard the Finsch's Conure as a sub-species of it. However, it is usually dealt with separately.

Finsch's Conure is almost unknown within the hobby of birdkeeping, and considering both the present international legislation and that to come, there will be no change in this situation. It is therefore very doubtful whether this species can be preserved for the birdkeeper. An observant enthusiast will still come across a pair every now and then; however, most keepers fail to recognize them as Finsch's Conures, often only seeing "another of those green South Americans with red on its head". As a result they do not get the appreciation which they deserve. This is a great pity, for there is a good chance that in a number of years this species will no longer be found in Western Europe.

Apart from Walsrode there have, so far, been no other reports of breeding success. Arndt has, in the meantime, had (unfertile) eggs. However, it is certainly not impossible that somewhere other young have been reared in an aviary at some time.

Birds which have the red extending to the eyes are almost identical in appearance to the *Aratinga wagleri minor*. However, the difference in size eliminates any possibility of confusion: the Finsch's measures 28cm (11in) and the *A. w. minor* 38cm (15in). What is more, the Finsch's greater under wing coverts are yellow.

Mutations

None known.

WAGLER'S CONURE - *Aratinga wagleri wagleri*

Sub-species

1. *Aratinga wagleri wagleri*   - Wagler's Conure
2. *Aratinga wagleri transilis*   - Peter's Conure
3. *Aratinga wagleri frontata*   - Cordilleras Conure
4. *Aratinga wagleri minor*   - Carriker's Conure

The forehead and skull of the nominate form are red, but it does not extend as far as the eye. There is a green stripe between the eye and the skull. The bend of wing is green, and the greater under wing coverts olive yellow. There are often red feathers on the neck and throat, and sometimes elsewhere on the body. Number 2 is darker green; the red on the skull extends less far but is deeper in colour. This subspecies is slightly smaller in size.
Number 3 is the largest; the red on the head extends to the eye. The bend of wing is red. Number 4 looks very similar to 3 but is slightly smaller and darker green. Moreover the legs of number 3 are flesh-coloured whereas those of number 4 have a grey tint.
Within this species, therefore, a clear division can be made according to the colour of the bend of wing: in 1 and 2 it is green, and in 3 and 4 red.

Origin of Name

Aratinga: shining parakeet.
Wagleri: after Dr Johann Georg Wagler (1800-1832), zoologist and director of the Zoological Museum in Munich.
Transilis: extend over, go beyond.
Frontata: with a mark on the forehead.
Minor: smaller.
Dutch: Wagleraratinga.
German: Kolumbiasittich.

Parents and Young

The plumage of adult birds is principally green, with varying amounts of red on the head, bend of wing and thighs. The iris of the nominate form is orangeish-brown, and the Cordilleras' is greyish. The amount of red on immature birds is reduced; number 1 has none at all. It is therefore difficult to determine the species of young green Aratingas.

Sizes and Weights

Length : Wagler's 36cm (14in), Peter's 34cm (13.5in), Cordilleras 40cm (16in), Carri-

9. The Cordilleras Conure is the largest Aratinga.

ker's 38cm (15in).
Weight: Cordilleras cock 265g, hen 225g; Carriker's 190g and 205g.
Ring size: Cordilleras 8mm (leg size cock 7mm, hen 6mm); Carriker's 7mm (leg size 5.5mm).

Habitat and Habits

Wagler's Conures are found at an altitude of between 800 and 2000 metres in the forests of the subtropical zone, in a wide band along the west coast of northern South America. Outside the breeding season they gather into large groups of about 300 birds and commute daily between their foraging and sleeping areas. Particularly in and around the roost trees there is a great deal of commotion, with plenty of squawking and bickering. They are not really tied to one area and often roam about. The climate influences their movements: if the higher regions become cold they come lower down, and they of course also follow the food sources which nature supplies.
They probably breed in colonies. Surprisingly they do not nest in holes in trees, but in cavities and cracks in almost inaccessible high rocks. It is known that this species has bred in one such site for at least 25 years.
The Peter's Conure forms even larger groups, which sometimes even total 1500 birds.
The Cordilleras Conure can be found on the mountain slopes of the Andes, but only in small numbers, as opposed to the Carriker's Conure, which like the other sub-species is fairly numerous. Next to nothing is known about the breeding habits of the three sub-species.

Diet

Wild: mainly tree fruits, nuts, berries and various seeds, which for the most part the birds find in the canopies of trees. To the great annoyance of the farmers they sometimes cause damage in orchards and grain fields.
Aviary:
1. sunflower seeds, millet, maize, wheat, pineapples, dried shrimps, lettuce, gurkins, snapdragons, chickweed, carrots and lots of fruit;
2. a seed mix of kardi, oats, a few sunflower seeds, hemp and buckwheat, lots of fruit, berries, vegetables and other greenery, twigs with fresh buds, minerals and vitamins.

Nesting Sites

As mentioned above this species nests in holes and cracks in rocks, often in inaccessible places. This is difficult to reproduce in an aviary. This poses no great problem as they will often use a box measuring about 50x20x20cm (19.5x8x8in) without any difficulty. The entrance hole should be 7cm (3in) in diameter. Inventive keepers could of course try making some sort of imitation cracked rock formation in their aviaries; it should be fairly easy to think of some construction, e.g. in concrete. This could be made horizontally and then stood up, in sections if necessary. Rainwater would have to be prevented from

entering into it. A construction such as this is often not only advantageous to the birds, but also gives a unique look to our often rather monotonously similar aviaries.

Breeding Process

The German aviculturist, Klössner, gives a clear report of his rearing of the nominate form, the Wagler's Conure. He acquired a pair in 1974 and housed them in an aviary measuring 6x3x2.5 metres (20x10x8ft), with an adjoining night shelter of 3x2 metres (10x6.5ft). On 27th, 28th and 30th May 1977 the hen laid three eggs in a nest-box measuring 50x25x25cm (19.5x10x10in), with a 10cm (4in) entrance hole. The first two chicks hatched on 20th June, followed by the third on the 22nd. Which gives an incubation time of 23 days. As the chicks were not fed, the attempt was unsuccessful that year. In 1978 the birds only began their pairing ritual in April (the year before at the end of February); the eggs were laid on 20th, 21st and 23rd June. One was infertile, the others hatched on 17th and 19th July. This indicates a longer incubation time of 26 days. As the hen died on 29th July, the chicks had to be reared by hand. As a rule three or four eggs are laid.

10. A close-up of a Cordilleras Conure (Photograph Birdpark Walsrode).

General Remarks

The nominate form was first bred in the USA in 1957, followed by Germany in 1978. In Britain, Chester Zoo has exhibited a colony of these birds since 1969; in 1978 they were on to their third generation of aviary-reared birds. Breeding results of the sub-species are not known, although their existence can certainly not be ruled out.

This species is far from common as an aviary bird. Pairs are occasionally seen in collections or at shows, and that is about it. Even then the keeper often does not know exactly which (sub-)species of the green Aratingas he has in his possession. The Peter's Conure has possibly never been imported, and the others only in very small numbers.

A disadvantage of these birds is their somewhat piercing voice, but apart from that they are lively aviary birds which are fairly easy to keep. As in the wild they live quite high in the mountains, they can be regarded as reasonably hardy. If they are able to spend the night in the nest-box a normal winter should not present any problems. An aviary three metres (10ft) long is the absolute minimum for these Aratingas, and the Cordilleras Conure must have four metres (13ft).

Considering the breeding habits in the wild it would be interesting to try breeding in a colony, in any case with the nominate form, possibly even using rock cavities. Arndt kept six Carriker's Conures in an aviary measuring 3x1x2 metres (10x6.5x3ft) for a considerable period, without any problems. However, it must be said that many parakeets can easily be housed together, and that problems only develop when they start breeding.

Mutations

None known.

MITRED CONURE - *Aratinga mitrata mitrata*

Sub-species

1. *Aratinga mitrata mitrata* - Mitred Conure
2. *Aratinga mitrata alticola* - Chapman's Mitred Conure

Number 2 has less red on its head: it is limited to a fairly narrow forehead stripe. The green plumage of this sub-species also has a blueish sheen.

Origin of Name

Aratinga: shining parakeet.
Mitrata: wearing a head band.
Alticola: living at a high altitude.
Dutch: roodmaskeraratinga.
German: Rotmaskensittich.

Parents and Young

The parents have an irregular red pattern on the head. However, the most important characteristic of this species is that the red completely surrounds the eye, which means that it could in fact only be confused with the Red-masked Conure.
After the Red-masked this is the species with the most red on the head. However, a comparison of the two photos shows that the Red-masked has considerably more red and, moreover, red edging on the wing, whereas the Mitred's wings are green.
Mitred Conures also often have the odd red feather scattered irregularly about the body. It seems that these do not always reappear in the same places after moulting. Hens may have less red on their heads, particularly on the cheeks, but this cannot really be considered a reliable means of sex identification. The iris is orangeish-yellow.
Young birds also display the odd red feather, but they are often fewer in number. Moreover, they have less red on the head, and it is usually limited to a broad forehead stripe. The eyes are of course still dark. The legs of young birds are noticeably darker than those of their parents.

Sizes and Weights

Length: Mitred 38cm (15in), Chapman's 37cm (14.5in).
Weight: hen 240g.
Ring size: 7mm (leg size 6mm).

11. The Mitred Conure's pattern of blotches is widely variable.

Habitat and Habits

The Mitred Conure replaces the Wagler's Conure in the southern Andes of central and southern Peru (where they occasionally occur together), and also occurs on the eastern slopes of the Andes through Bolivia into northwest Argentina. The two species are closely related. The Mitred prefers relatively dry mountain slopes and valleys, and usually forages in areas of scattered woodland. When seen flying, they are often covering large distances at a great height above half open countryside. When breeding, it appears that unlike the Wagler's, they are not dependent upon cracks and cavities in rocks. Outside the breeding season Mitreds can form flocks of more than a thousand birds at altitudes between 1000 and 2500 metres.

Number 2 has a limited distribution range round the Peruvian town of Cuzco, at an altitude of 3400 metres. It is clearly a mountain bird, which spends the night in large flocks in the rainforest. This species is locally common and is fairly numerous in areas which have been under cultivation for some time.

Diet

Wild: seeds, fruits and berries.
Aviary:
1. sunflower seeds and oats, a seed mix for large parakeets, assorted germinated seeds, lots of fruit and greenfood, fresh twigs and a self-made egg mix;
2. a parrot seed mix as a basis, with in addition, germinated sunflower seeds, lots of fruit and carrots, lettuce, chickweed, rosehips, soft maize and fresh twigs. An amount of this can be deep-frozen for use out of season. A vitamin preparation is added to the drinking water once a week. During the breeding season an egg mix is also given, along with boiled chicken egg, eggshells and a calcium preparation;
3. a mixture of concentrated sheep pellets and sunflower seeds in a ratio of 70 - 30, with in addition fruit, bread soaked in milk and honey, and germinated sunflower seeds.

Nesting Sites

Only one case of breeding has been reported in the wild. The nest was in a hollow tree about ten metres above the ground, and contained two eggs. The nest cavity was fairly open and the entrance relatively large.

Reports about breeding in aviaries give very little information about the nesting site. A British keeper reports success in a cube with 30cm (12in) sides. A box 40 to 50cm (16 to 19.5in) tall, with a bottom surface measuring 15 to 20cm (6 to 8in) square is probably better, or a natural site with a diameter of about 20cm (8in). The entrance hole should be about 7 or 8cm (3in).

Breeding Process

The first breeding success in Germany was in 1982. Three eggs were laid weighing 14g

and measuring 28x34mm. The hen started sitting after the first egg. The incubation time is reported to have been 26 days, but it is impossible to be sure of this as only one of the three eggs hatched, and this was not necessarily the first. The chick emerged from the box after 59 days, and weighed then 220g.

The German aviculturist, Bauer, bred this species in 1984. He purchased two birds on 17th February and 4th March and on 5th, 7th and 8th May of the same year, three eggs were laid in a natural log in the shelter; they were fertile but failed to hatch. A second clutch was laid on 25th, 27th and 29th June, and the chicks hatched on 18th, 20th and 22nd July, which indicates that the hen started incubating immediately after laying the first egg. The incubation time can be easily worked out: 23 days. The chicks left the nest on the 66th day. At that time they had only a red stripe on the forehead.

A British keeper put two eggs in an incubator. They both measured 26.2x32.5mm. The newly-hatched chicks weighed 8.27 and 8.38 grams. Their eyes opened after about two weeks, and they were ringed on the 23rd day. At that age, red feathers were already visible on the head. They were independant at an age of nine weeks.

General Remarks

At Loro Parque in Tenerife they had success with this species in 1981, Germany followed in 1982 and Switzerland in 1986.

In aviaries these birds are avid bathers. Generally speaking they remain fairly shy for a considerable time. They are hardy birds which feel perfectly happy in low temperatures.

Although they are not unknown as aviary birds, you only come across them occasionally. A sufficient breeding stock has probably not yet been built up and the future of this species as an aviary bird is not yet guaranteed. However, it is not too late do something about this.

Mutations

None known.

12. A close-up of a Mitred Conure.

13. On leaving the box young Red-masked Conures have a completely green head; the young bird in the photo is slightly older

RED-MASKED CONURE - *Aratinga erythrogenys*

Sub-species

None.

Origin of Name

Aratinga: shining parakeet.
Erythrogenys: with red cheeks.
Dutch: Guayaquilparkiet.
German: Guayaquil-Sittich.

Parents and Young

Males sometimes possess a slightly larger head and bill, and are a little more heavily-built. However, these differences are not so reliable that they can be entirely depended upon. An added difficulty is that both sexes have the same plumage. The amount of red on the head varies, but so far this does bear any relation to the sex of the bird. One advantage of this is that it is fairly easy to tell individuals apart. I have a bird in which the red extends as far as the neck, and so far I have seen only one other similarly marked bird. Besides the head, the bends of wing and thighs are also red. The iris is orangeish-yellow.
My first chicks left the box with an almost completely green plumage; they only had a brown forehead band and red under wing coverts. When it was five months old red feathers began to come through on the crown of the first young. A good month later virtually the entire forehead and crown was red. Only then did the odd red feather begin to appear on the second young's cheeks.

Sizes and Weights

Length: 33cm (13in).
Weight: cock 185g, hen 165g.
Ring size: 7mm.

Habitat and Habits

This species occurs in a small area in the dry tropical zone of southwest Ecuador and northwest Peru, where it seems to lead a somewhat nomadic existence and is therefore not

always to be found in the same place. Within this limited distribution range the Red-masked Conure seems to be fairly common, and lives in a variety of habitats ranging from deciduous woods through areas of dry scrubland to sparsely vegetated open country, and even in villages and towns. It is principally a bird of the dry and low-lying regions up to 500 metres. The closely-related Wagler's Conure lives at a higher altitude.

The Red-maskeds roam about in extremely large flocks outside the breeding season. During the breeding season the various pairs isolate themselves from the group.

Diet

Wild: strangely enough it is not exactly known what these birds eat in the wild, but the menu probably consists of fruits, berries, nuts, leaves, seeds etc.
Aviary:
1. a good seed mix for parakeets with in addition a little oats, wheat and barley. Also a daily helping of egg mix and plenty of fruit and greenfood;
2. sunflower seeds, hemp, white millet, rice and maize; also sweet apple, all sorts of greenfood (including parsley) and twigs with fresh buds.

These Aratingas are not very fussy. They consume completely the seed mix which is placed before them. They are very partial to cereals grown here such as wheat, oats and barley and these can be given in moderation. As they are fairly inquisitive they actually try anything which is served them.

Nesting Sites

It seems that their natural nesting sites have yet to be discovered, so that next to nothing is known about their breeding habits. This is quite remarkable for a species which can even be found in towns and villages.

My own birds nest in a box 53cm (21in) tall whose outside measurements are 28x26cm (11x10in), which is hung in the night shelter. Although I had made a 7cm (3in) entrance hole in it, they have gnawed it out to 11.5cm (4.5in).

Results have also been achieved with a box 65cm (25.5in) tall with a floor area of 30x35cm (12x14in) and a 9cm (3.5in) entrance hole, with one of 45x30x30cm (18x12x12-in), and with one of 35x20x20cm (14x8x8in).

Breeding Process

On 25th July 1982, I found three eggs in the box of birds which had been imported in 1979. They were warm so incubation had therefore already started; several days later they proved to be fertile. Nest control was made easy for me; the hen was still rather shy and left the box as soon as I opened the door of the aviary. She did this during the entire incubation period; but despite this she carried out her task admirably. On 12th August the first egg was found to be pipped and on the 13th the chick first saw the light of day. The second chick followed on the 16th. The eyes of the oldest began to open on 25th August. On 26th and 30th August, the 13th and 14th day, the chicks received a 6mm ring, which afterwards proved to be on the small side. They left the box on 10th and 11th October at an age of 58 and 56 days. By another birdkeeper this occurred on the 55th day. Although figures of 23, 26, 28 days and a month are found mentioned in reports, the average incubation period can be fixed at 24 to 25 days. A clutch consists of three or four eggs. They are often laid rather late in the season, sometimes even in July.

General Remarks

As far as is known, these Aratingas first arrived in Europe in 1854, when they were brought to London. The British aviculturist, Shore-Baily, appears to have bred them for the first time in 1925. In 1969 two young were reared in Chester Zoo, and in the same year four young were also reared in Denmark.

I myself bought four birds from an importer in 1979 which were not yet fully-fledged, three of which finally survived. For some years they were allowed to gradually adjust to the transition from life in the wild to life in an aviary in spacious accomodation. They proved to be hardy birds which were easily able to withstand the Dutch climate. They were not fussy with regard to food and got on well together. However, in the summer of 1982 all this changed. One of the three was clearly no longer being accepted by the other two. They would, for example, continually chase it, and it was no longer allowed to sit on the same perch. At this point I removed it from the aviary. It soon transpired that this had been the correct decision, for a good week after the separation I found on inspection three eggs in the box. Two healthy young birds developed from these eggs.

14. A red-masked Conure cock with an exceptional amount of read on its head; it differs greatly from other birds of the same species.

When I came across an exceptionally fine example of this species at a bird market purely by accident (the above mentioned bird with a great deal of red on its head), I decided to purchase it. An endoscopic investigation some time later showed the bird to be a cock. After much deliberation I sold the cock of my breeding pair and replaced it with the new one. The two have been together now for a number of years and get along fine. They feed each other and mate, but so far they have not produced any eggs. Which once again only goes to show that it is not very sensible to split up good breeding pairs.

Of the species of green Aratingas this is the one with most red on its head and it is undoubtably also the most beautiful. However, there are very few keepers who make a serious attempt to breed them. Although they have been imported regularly for a number of years in fairly small numbers they are still not widely found in collections.

Mutations

None known.

15. The White-eyed Conure is one of the least strikingly coloured Aratingas.

WHITE-EYED CONURE - *Aratinga leucophthalmus leucophthalmus*

Sub-species

1. *Aratinga leucophthlamus leucophthalmus*   - White-eyed Conure
2. *Aratinga leucophthalmus callogenys*   - Ecuadorian White-eyed Conure
3. *Aratinga leucophthalmus propinquus*   - White-eyed Conure
4. *Aratinga leucophthalmus nicefori*   - Nicéforo's White-eyed Conure

Generally speaking it is fairly difficult to distinguish between the different sub-species with any degree of reliability, all the more because their distribution ranges overlap and it is possible that crosses exist. However, the following differences are mentioned in a cautious attempt, although an added problem is that this information is only really useful if it is possible to compare different subspecies directly. And this situation will occur only very rarely. The only consolation is that most of the birds found in Europe belong to number 1.

At first sight the nominate form appears almost entirely green, with the odd red feather scattered about the neck and throat. The bend of wing is also red.

Number 2 is slightly larger and the bill is noticeably heavier; the green is a shade darker.

There is some uncertainty regarding number 3; the only difference with number 1 is that it is slightly larger. It may even be that it is not a separate sub-species.

Number 4 is also similar to the nominate form, but the plumage is more yellowish-green and it has a narrow forehead stripe. However, uncertainty exists here too, as only one bird (caught in 1946) is known.

Origin of Name

Aratinga: shining parakeet.
Leucophthalmus: white eyes.
Callogenys: beautiful cheeks.
Propinquus: similar to, related to.
Nicefori: after the Frenchman Nicéforo Maria (1888-1980), who lived in Columbia from 1908 onwards and became one of South America's most famous zoologists.
Dutch: witoogaratinga.
German: Pavuasittich, Weissauggensittich.

Parents and Young

Mature birds are almost entirely green with only a scattering of red feathers on the head, neck and throat. The outside lesser under wing coverts are red, causing the red bend of wings, the greater are yellow. The iris is orange. There is no difference to be found in the plumage of the sexes. The only (not particularly reliable) difference is in the size of the

head and bill.
Young birds are very similar in appearance to their parents but do not have red feathers. The wing edge is yellowish-green.

Sizes and Weights

Length: White-eyed 32cm (12.5in), Ecuadorian 34cm (13.5in), *propinquus* 35cm (14in), Nicéforo's 34cm (13.5in).
Weight: 130g.
Ring size: 7mm (leg size 6.2mm).

Habitat and Habits

The White-eyed Conure is the species within this genus with the largest distribution range, about half of South America. As a result there is a natural variation in habitat and habits. In general they prefer the edges of woods, and areas with low vegetation and groups or rows of trees. In the wetter regions (the tropical rainforests and marshlands) they tend to live along water courses. This species does not venture high into the mountains. In Bolivia it is found up to 1800 metres, but this is not reached elsewhere.
These common Aratingas are normally seen in pairs, family groups or small flocks of ten to thirty birds. Although in Peru one night roost has been found to which at least 500 birds return every evening. When they arrive the noise is deafening, as they constantly squabble for the best spots. Peace only returns when it is completely dark. In Venezuela they have been found together with Sharp-tailed Conures. They are excellent fliers, and can cover great distances when commuting between their foraging and roosting places.
The population is stable; this is probably due, on the one hand to the size of the area in which they occur, and on the other to the fact that as a result of their rather unimpressive plumage they are of little interest to birdkeepers.

Diet

Wild: the menu consists of berries, fruits, seeds, nuts, blossoms, greenery, and insects and their larvae, while in some regions they make grateful use of cultivated sugar cane. In the stomach of two birds caught in Brazil were found 180 and 600 grass seeds; in the stomach of a third bird there were 50 seeds, and the remains of fruit and insects.
Aviary:
1. sunflower seeds, hemp, millet, canary seed, apples and pears and occasionally other fruit, also bread and milk;
2. sunflower seeds, a budgerigar mix, hemp, apples, pears and the odd piece of orange or banana.

Nesting Sites

It seems that White-eyed Conures retreat into the woods during the breeding season, only to emerge again once the chicks have left the nest. They nest in holes in the trunks and large branches of trees.

Birdkeepers have achieved breeding successes in boxes measuring 45x30x30cm (18x12x-12in).

Breeding Process

Clutches consist of three to four eggs. Information about breeding successes is limited, and at times seems rather odd, for example the length of time before the young leave the box is given in three separate cases as 42, 65 and 76 days. Of these 65 seems to me the most reliable as in that case the breeder gives the exact dates. However, the diet given was not very good so it may have taken longer than usual. The incubation time is probably comparable to that of the other green species.

General Remarks

The first breeding success was in the USA in 1934, when young were twice reared. Australia followed in 1937, and South Africa in 1973. The first success in Europe was not reported until 1975 in Britain. In the Netherlands an award for an exceptional breeding success was given in 1983.

Once they are acclimatized they are hardy birds which require no special care, as long as they receive a suitable diet. They can become fairly aggressive when they have eggs and chicks; other than that they become quite tame in time. They sometimes produce rather a lot of noise.

Mutations

Nelson Kawall in the Brazilian city of Sao Paulo possesses two cinnamon birds.

HISPANIOLAN CONURE - *Aratinga chloroptera chloroptera*

Sub-species

1. *Aratinga chloroptera chloroptera*   - Hispaniolan Conure
2. *Aratinga chloroptera maugei*   - Mauge's Conure

Number 2 is now extinct and was last seen in 1912. It was almost identical to number 1 but had more red on the underside of the wings. The mention of two sub-species is therefore of no practical value, all the more as number 2 is only known as three museum exhibits.

Origin of Name

Aratinga: shining parakeet.
Chloroptera: green wing.
Maugei: René Maugé de Cely was the most important naturalist during the French captain Nicolas Baudin's expeditions to the West Indies (1786-1788) and Australia (1800-1804).
Hispaniola: an island in Central America, divided into Haiti and the Dominican Republic.
Dutch: Hispanolaparkiet.
German: Haitisittich.

Parents and Young

Adult birds are almost completely green in appearance; some have a few red feathers on the head. The outside under wing coverts, and the bend of wing are red. The iris is reddish-brown, and the bird has a striking white eye ring.
Young birds look similar to their parents, though they have noticeably less red.

Sizes and Weights

Length: 29cm (11.5in).
Weight: 135g to 170g.
Ring size: 7mm.

Habitat and Habits

This species occurs throughout the island of Hispaniola and in all the habitats found on it; however, they are clearly more at home in mountainous regions. They are not especially shy but do make rather a lot of noise. They commute daily between their foraging and

16. The Hispaniolan Conure can only be found on the island which is it named after (Photograph Birdpark Walsrode).

roosting places.
They mainly search for food in the treetops, sometimes in groups of up to 100 busy birds. The pairs are maintained within the group, and they isolate themselves from it during breeding.

Diet

Wild: these birds feed on many sorts of seed, fruits, and berries, nuts and probably blossom and leaves.
Aviary: on Grand Canary (Palmitos Park) they are particularly fond of fresh maize and millet sprays, along with a mixture of small germinated seeds (canary seed, hemp and oats). Their normal diet consists of a mixture of germinated sunflower seeds and boiled maize, with a daily supplement of such things as peas, carrots, boiled French beans and boiled rice. Calcium carbonate is added a few times a week, and when there are chicks a calcium preparation is added to the drinking water. Finally they receive sliced fruit.

Nesting Sites

In the wild they are not particularly fussy about the choice of a temporary home in which to raise their young. Nests have been found in, among other places, deserted termite hills, woodpecker holes and dead trees, up to a height of 25 metres above ground.
Should you come into possession of these birds, offer them a box measuring about 50x20x20cm (19.5x8x8in) with an entrance hole 7cm (3in) in diameter.

Breeding Process

Not much is known about the reproduction of this species. Reports from the island of Hispaniola mention clutches of maximum seven eggs, which compared to other Aratingas is quite a lot. It seems that the average number in the wild is three to five.
Rosemary Low mentions that they were first bred in Europe in the Palmitos Park in Grand Canary. This was in 1986, when two young were reared from a clutch of four eggs. Since then the parents have produced one or two young each year. In 1990 a second pair bred successfully.
Roemary Low gives more detailed information about the breeding process of one pair in 1989 and one in 1990. On 5th May 1989 an egg was found in the box; on 10th May there were two. They hatched on 2nd and 5th June, which indicates an incubation time of 26 days for the second egg. Both chicks left the box on 30th July, 58 and 55 days after hatching.
On 15th November an egg was again found, and on the 17th a second. They hatched on 10th and 13th December, which once again means an incubation time of 26 days for the second egg. They were ringed on the 16th and 15th day, and they left the box on the 60th and 57th day.
In 1990 eggs were laid on 5th and 8th April. The first chick hatched on 2th May, the second on 6th May. They left the nest on 26th and 29th June.

General Remarks

There is mention of a breeding success in the USA in 1936. To my knowledge in Europe this species is only found in Walsrode Bird Park in Germany and in Palmitos Park in Grand Canary, where they were first bred in 1986.
At first sight the Hispaniolan looks very similar to the White-eyed Conure. The most obvious difference can be found in the colour of the under wing coverts, which in the Hispaniolan are much redder and lack any yellow. On direct comparison the eye ring of the Hispaniolan is also noticeably whiter, and this species has fewer red feathers on its head.

Mutations

None known.

17. The Cuban Conure is a less common aviary bird than the Cuban Amazon.

CUBAN CONURE - *Aratinga euops*

Sub-species

None.

Origin of Name

Aratinga: shining parakeet.
Euops: nice appearance.
Dutch: Cubaparkiet.
German: Kubasittich.

Parents and Young

The plumage is mainly green, but appears speckled due to the red feathers which occur over the entire body. The belly and breast have a more yellowish-green colour. The greater under wing coverts are olive yellow, and the lesser are red. The bend of wing is also often red. The iris is yellow.
Young birds have less red under the wings and the wing edge is green.

Sizes and Weights

Length: 26cm (10in).
Weight: approximately 100g.
Ring size: 6mm.

Habitat and Habits

This species only occurs on the island of Cuba. It has a clear preference for rugged inaccessible woodland areas, but is occasionally found in more open country. As a result of forest clearance, it has therefore become scarce in large areas of the island. They roam about the remaining country in groups of about thirty birds, sometimes in the company of Cuban Amazons.
Their plumage is an excellent camouflage and they are difficult to spot. They are easier to track down by their call. It seems that they are not particularly shy.
In 1987 the German aviculturist, Weissweiler, travelled about the island for days without seeing a single bird, not even a caged one. Neither could the local inhabitants give him any information, except that some farmers suspected that they were still to be found on the island in the Trinidad mountains.

Diet

Wild: seeds, leaf buds, nuts, blossom, fruits and berries. Families and other larger groups forage for food in the treetops.
Aviary: nothing known, but should be related to the above, and to the eating habits of other green Aratingas.

Nesting Sites

It appears that they prefer to nest in the deserted holes of the native green woodpecker. These are generally found in palm trees and termite hills. However, other tree holes are also used.

Breeding Process

Between two and five eggs are laid in the above mentioned nests. The first breeding success was in East Berlin where four birds were being kept in one aviary. One of the hens laid four eggs, which hatched after 23 days. The oldest chick left the box after 48 days.
The German aviculturist, Fuss, from Munich, acquired this species by accident at the end of the seventies. He came across the first bird being kept in a cage as a pet. Some time later he was visiting an importer when to his surprise he discovered another one among a shipment of White-eyed Conures. As the importer was not aware of this he was able to purchase it for the same price as the White-eyed. Fuss was lucky because it turned out to be a pair! The birds had been together for a number of years when in March 1986 eggs were laid and one young was produced. In the autumn of that year a further four eggs were laid. They were fertile but nothing came of it as the hen deserted them.
In 1987 the birds missed a year, but in 1988 they were successful again and produced two young.

General Remarks

The first known breeding success was in East Berlin in 1967.
In Western Europe the Cuban Conure is a rarity and is virtually unknown. I myself have only seen it in the Loro Parque in Tenerife.
In Eastern Europe the situation is slightly better, as a result of the (former) good relationship with Cuba. In the past few years they have been bred there fairly regularly. However, reports of these cases are not available.

Mutations

None known.

18. The golden-fronted Conure differs from the nominate form by its completely green rump and the lack of yellow on its cheeks.

GOLDEN-HEADED CONURE - *Aratinga auricapilla auricapilla*

Sub-species

1. *Aratinga auricapilla auricapilla* - Golden-headed Conure
2. *Aratinga auricapilla aurifrons* - Golden-fronted Conure

These two look very similar. The first difference can be found in the lores, the area between the eye and the lower mandible. The feathers of number 1 are here principally yellow to brownish-red, number 2's are green. A second, and more reliable, difference is the feathers of the rump: number 1's are edged with reddish-brown, whereas 2's are pure green. Finally number 2 often has more yellow on the crown. When breeding it makes great sense to keep both sub-species pure. There are now enough birds available to make this possible. However, it is not always easy as their distribution ranges border on each other, so that even in the wild crosses occur.

Origin of Name

Aratinga: shining parakeet.
Auricapilla: golden hair.
Aurifrons: golden forehead.
Dutch: goudkaparatinga.
German: Goldscheitelsittich, Goldkappensittich.

Parents and Young

Adult birds display splendid contrasting colours; yellow and reddish-brown against a dark green background. The shades of the yellow and reddish-brown on both the head and the belly, where the tints are darker, can vary considerably. This is no indication of the sex and this cannot be determined from the plumage. Young birds are very similar in appearance to their parents; the most important difference being the reddish-brown and yellow colouring, which is less intense and not so widespread. They are sexually mature after two years.

Sizes and Weights

Length: 30cm (12in).
Weight: 140, 143 and 158g.
Ring size: 6mm.

Habitat and Habits

The presence of woods is essential for this species; it prefers the fringes of woods and small clearings. As southeastern Brazil is a region which is the victim of large-scale forest clearance, the numbers of this species have been decreasing rapidly for some years. It is dependent upon woodlands for its existence, and is not able to move to other parts of the South American continent.
As a result its distribution is somewhat irregular, and particularly in the southern region the situation is very bad. It is therefore found in increasingly small numbers, and this process is still going on.
The same applies to at least eleven other species of parrotlikes which occur in the same region and are reliant upon woodland.
Generally speaking Golden-headeds are quieter by nature than their close relatives the Jendaya Conures, which are noisier and bolder.

Diet

Wild: nothing is known, but it is probably comparable to that of the Jendaya Conure.
Aviary:
1. a seed mix for larger parakeets, white and striped sunflower seeds, maize, peanuts, apples, rosehips, greenfood and willow twigs; supplemented in the breeding season with egg mix, soaked maize, oatmeal and baby food;
2. during rearing, in addition to a seed mix, fruit, lettuce, sunflower seeds and a rearing mix for chickens;
3. germinated sunflower seeds and wheat; a seed mix of hemp, paddy rice and buckwheat; apples and rowan berries, and twice a week a multivitamin preparation in the drinking water;
4. a seed mix for parakeets with, depending on the time of year, a good egg or standard mix; fruit all year round, rosehips, twigs and soaked white bread;
5. a mix with a third part sunflower seeds and consisting further of various sorts of millet, hemp, buckwheat, paddy rice, linseed, wheat and oats; apples, pears and oranges; and finally carrots, germinated sunflower seeds, rosehips and maize cobs. During the breeding season a hard-boiled egg daily, mixed with egg mix and a little calcium.

Nesting Sites

Birdkeepers have reared young in boxes measuring, among others: 50x30x30cm (19.5x 12x12in), with an 8cm (3in) entrance hole (nesting material: a mixture of peat and wood shavings); 53x27x27cm (21x10.5x10.5in), with a 9cm (3.5in) entrance hole; and 100x 25x25cm (40x10x10in).
A British aviculturist who has bred these birds, noticed that he got better results with Aratingas if they had fairly tall boxes. He uses ones measuring 106x23x16cm (42x9x6in), and rears young without difficulty.

Breeding Process

Three to five eggs are laid, and the incubation time is about 25 days. Immediately after hatching the chicks weigh seven or eight grams. The hen remains in the box for about two weeks and is regularly fed by the cock. At an age of about fourteen days the chicks reach a peak weight of about 150-160 grams. Shortly before leaving the nest this decreases slightly.
The eyes open after about eleven days. This is also the time to ring the chicks. The bill begins to darken after about three weeks. In a number of registered cases the chicks left the box after 50, 54, 56 and 58 days.
Two clutches in one year is a possibility. The eggs are often laid quite far into the season; it may well be June. A German keeper nearly always has chicks in the autumn.

General Remarks

The first breeding success was probably in 1930, when a Californian keeper had young. The first report from Britain was in 1979; the keeper in question reared no fewer than 23 young from two pairs in three years. In the Netherlands they were probably first bred in 1976. However, breeding awards were only registered in 1980 (*A.a.auricapilla*) and 1983 (*A. a. aurifrons*).

19. The Golden-headed Conure is under threat as a result of population growth and forest clearance.

The birds present in the Netherlands are virtually all the result of imports made during the mid-seventies when the first shipments to reach Western Europe arrived. Before that time they were unknown as an aviary bird, so their history here is very short. As Brazil has now imposed a total export ban there is no reason to hope that those shipments will be repeated.

The Golden-headed is closely related to the Jendaya Conure and the Sun Conure. Some experts even regard them as the same species.

Mutations

Two lutino mutations were known to exist in Sao Paulo in 1974. Unfortunately, as nothing else has been heard about them, it is to be wondered if they are still alive.

20. The Jendaya is closely related to the Golden-headed and the Sun Conure.

JENDAYA CONURE - *Aratinga jandaya*

Sub-species

There are no sub-species. However, this species does have one unusual characteristic, namely that some birds have a white eye ring and others have a black one. According to Low, the colouring of the form with the black eye ring is more intense, resembling more closely that of the Sun Conure.
The plumage of birds with a white eye ring is supposed to contain yellow instead of the reddish-orange. However, according to my own observations this is not always the case. Arndt considers the possibility of there being two sub-species, the difference between which is only slight.
It is a pity that nothing is known about the situation in the wild in South America. It would be interesting to know, for example: whether the two forms inhabit separate regions, or have overlapping distribution ranges; whether the range of one is closer to that of the Sun Conure than the other; if the two forms mate with each other, or that they remain strictly separate during reproduction.
Future investigations will hopefully answer these questions. I have the impression that birds with a black eye ring are more common here. In any case it does not seem a good idea to me to cross the two forms.

Origin of Name

Aratinga: shining parakeet.
Jandaya: Indian name for this bird.
Dutch: jendayaparkiet.
German: Jendaya-Sittich.

Parents and Young

There is no difference between the plumage of cocks and hens. That is not to say that all birds look alike. Particularly the colour of the belly can be variable; ranging from a slightly orangeish-yellow to orangeish-red. It is not impossible that this has to do with the region birds comes from; in general the closer to the equator birds live, the more intense their colours become. The difference in eye ring has been mentioned above. There may be some link between the two.
The yellow and orangeish-red parts of the body are clearly paler in young birds. Moreover, quite a few small green feathers are visible particularly on the head, neck and throat. They become sexually mature at the end of the second year, although I came across a case of a one-year-old bird producing young. My own second generation birds only produced eggs after three years. However, that was only to be expected as after more than two years it transpired that my 'pair' were both hens.

Sizes and Weights

Length: 30cm (12in).
Weight: cock 125g, hen 140g. Two of my own young, which were sexed after two years weighed 112g and 126g (both hens).
Ring size: 6mm.

Habitat and Habits

This species inhabits the so-called caatinga region (lots of dense, often thorny, low trees and bushes), and the Brazilian mountain regions. Nowadays they also occur in open country which was at one time covered with rainforest. The entire distribution range has a tropical savanna climate; hot and wet with short dry periods.
The rainy season lasts for seven to nine and a half weeks, interrupted by two dry spells. The temperature can vary enormously within a twenty-four hour period; from 40°C during the day to 10°C at night.
It will be clear that this species is very adaptable and is not overparticular with regard to its surroundings. It is indeed common within its range, although numbers do vary in different areas. It is not found in the dry areas of scrub.
On the whole it seems that the Jendaya Conure is rather on the increase than the decrease. In coastal areas it is even one of the commonest birds, and travels about in flocks of ten to twenty birds, preferably remaining in coconut palms. It is a lively guest in any aviary, with an inquisitive and investigative nature. However, they can also be rather noisy. This is a characteristic which does not appeal to every birdkeeper, even if only because of the possible nuisance caused to the neighbours.

Diet

Wild: seeds, berries and other fruits and a fair amount of greenery.
Aviary:
1. standard seed mix for larger parakeets, supplemented with egg mix, soaked white bread, fruit, greenfood, fresh weeds and willow twigs;
2. sunflower seeds and oats, as well as egg mix mixed with dog chow and a little chickweed; also willow twigs.

Nesting Sites

Nothing is known about sites in the wild.
Birdkeepers have recorded successes using both a box measuring 55x30x30cm (22x12x 12in) with an 8cm (3in) entrance hole, and a 130cm (51in) tall natural site. I personally breed them in a box measuring 53x24x27cm (21x9.5x10.5in) with an entrance hole 7cm (3in) in diameter.

Breeding Process

Jendayas lay three to four eggs; breeding reports give an average incubation time of 25 days. Chicks can be ringed on about the tenth day. The number of days before the chicks leave the box has been registered as 50, 56, 58 and 66. Eggs have been laid in February, May, June and August. Two clutches a year sometimes occur, but is not normal.
After a couple of difficult years with my own birds I first had chicks in 1985. I eventually had them sexed, and when in that spring I finally had a good pair I had two immediate successes, and they reared a total of six young from seven eggs. The first clutch was laid in April. In the years following they have restricted themselves to a single clutch in May, which always consists of four eggs. The incubation time varies between 23 and 25 days, and I can ring the chicks after ten days at the earliest. The number of days before they leave the box also varies: the earliest was after 55 days, the latest after 63.
Strangely enough this reliable pair skipped a year in 1988, when they failed to produce even a single egg. However, in 1989 they picked up where they had left off.
Breeding in colonies has been attempted a number of times, but on the whole the results are less good than with separate pairs. The birds probably disturb each other too much, causing so many distractions that they get insufficient peace to be able to breed. After all, it is known that in the wild breeding pairs of Aratingas isolate themselves from the group and withdraw.

General Remarks

As far as is known, the first Jendayas in Europe could be viewed in a zoo in 1854; in 1869 four birds arrived at London Zoo. The first breeding success dates from as early as 1890 (Britain). Further successes are: West Germany 1926, USA 1932, Portugal 1957, Sweden 1960 and Denmark 1964. In the Netherlands the first bird to be bred was exhibited at the 1979 Dutch Parrot Society show, and it received an award in 1980; however, earlier successes had undoubtably been achieved.
This species is no longer imported as it only occurs in Brazil, which has had an export ban in effect for some time.
The degree of destructive chewing, which is feared by many, varies greatly. My breeding pair hardly does it at all, but a pair of young birds which had been housed separately destroyed a box of rough planks so completely that no bird could have bred in it. Afterwards I hung up a new plywood box and the chewing suddenly stopped. They moved into the box immediately, and several months later nested in it.
It is possible to allow Jendayas out of the aviary. The Dutch aviculturist, Rumonder, regularly lets a few out and they return to their homes without fail.

Mutations

None known.

SUN CONURE - *Aratinga solstitialis*

Sub-species

None. It has already been mentioned in the description of the Golden-headed Conure that the Sun Conure is closely related to both that species and the Jendaya Conure. Some systematists even go so far as to consider them a single species. The discussion is still going on. However, such a discussion is of little interest to the average birdkeeper, although he will have to agree that the three are more or less closely related.

Origin of Name

Aratinga: shining parakeet.
Solstitialis: belonging to the sun.
Dutch: zonparkiet.
German: Sonnensittich.

Parents and Young

Mature birds display various shades from yellow to orangeish-red on head, belly and breast, and the wings are also mainly yellow. The remainder is green, although the primary flight feathers tend to blue. In my experience the naked eye ring is always whiteish, although Jim Hayward states that it can also be dark (see the Jendaya). He suggests the possibility that the skin pigment is affected by the diet. However, so far I have never detected any change in colour.
Little or no external difference can be found between the sexes. Some people maintain that hens have more green on the wings, and that cocks have more orangeish-red in the yellow. However, this is not one hundred per cent reliable.
Most young birds have almost entirely green wings with only a yellow top leading edge. However, some do have a fair amount of yellow on the wings on leaving the box. Belly and breast are greenish-yellow, the head is already a deeper yellow and orange but a green sheen is still visible. The colours are generally paler. They develop their full adult plumage at an age of eighteen months. This coincides with sexual maturity, which means they are able to produce young after two years.

Sizes and Weights

Length: 30cm (12in).
Weight: approximately 130g.
Ring size: 6mm.

21. It is not difficult to see why this species received the name Sun Conure.

Habitat and Habits

The Sun Conure inhabits savanna, palm woods, and open light woods up to an altitude of 1200 metres in a warm and damp climate. Their distribution varies. In some places they occur in large numbers, usually in small groups, but sometimes in very large flocks. They are then very noisy. They remain within their distribution range north of the Amazon and are therefore widely separated from the Jendaya Conure. Otherwise little is known about their habits in the wild.

Diet

Wild: just as the other Aratingas their diet consists mainly of fruit, nuts, seeds, and insects and their larvae. They are often found in large numbers in trees bearing ripe fruits, where they create a great commotion.
Aviary:
1. sunflower seeds, oats, wheat, millet and millet sprays; vegetables and fruit; young birds receive hard-boiled eggs and biscuits and a soft feed consisting of boiled rice, eggs, meat, milk powder, dextrose, and a mineral and vitamin preparation; finally fresh twigs;
2. a wide variety of fruits and vegetables, including maize cobs, apples, oranges, carrots, beetroots and grapes. Also seeding grasses and various other seeds, also in germinated form;
3. sunflower seeds, peanuts, maize, pineapple, a seed mix for large parakeets, millet sprays, apples and carrots; also egg mix, germinated seeds, chickweed, snapdragons, half ripe grass seeds and maize cobs;
4. seed mix for large parakeets, large amounts of fruit, maize cobs, greenstuff, half ripe weed seeds, blossom, bark, boiled whitefish, dried shrimps and mealworms;
5. seed mix for large parakeets supplemented with sunflower seeds, paddy rice and unhusked oats; as rearing food dried and ground bread, Protifar, boiled egg and rosehip juice; greenfood and fruit.

Nesting Sites

Nesting sites are found in holes in palms and other trees. A German keeper gave these birds the choice between a box measuring 45x30x30cm (18x12x12in) and a natural site 40cm (16in) in diameter and 80cm (31.5in) tall. Their preference went to the first. However, when the natural site was replaced with another 120cm (47in) tall and 45cm (18in) in diameter they moved into it immediately and reared young there.
Successes have also been recorded in boxes measuring: 40x25x25cm (16x10x10in) with an 8cm (3in) entrance hole; 46x18x18cm (18x7x7in) with a 7cm entrance hole; and in a 150cm (60in) tall natural site.

Breeding Process

As the breeding season approaches the birds mate regularly and the hen spends more time

in the box during the day. Particularly the latter is a fairly reliable sign; mating in itself does not indicate anything, and can take place without any eggs being produced. In such cases it often serves to strenghen the pair bond. Clutches usually contain four to six eggs, and incubation usually starts after the second egg. They are normally laid in the period February to July. According to reports from reliable breeders the incubation time is 23 to 25 days, about the same as I myself have noted for Jendayas.

The young leave the box at an age of 48 to 56 days. They return to the box at night to sleep. This does represent a risk if a second clutch has been laid, as the eggs can become damaged. It is better to remove the young as soon as they are independent.

The male generally spends a good deal of time in the box with the hen before she has laid the eggs and they can be clearly heard chewing away at the inside in order to make a soft bed of splinters for the eggs.

General Remarks

Linnaeus described this species in 1754 and gave it its scientific name, but it had already been named by the German Johann Leonhard Frisch in 1733.

The Sun Conure was on exhibition in Berlin Zoo in 1845, and in London Zoo in 1862. The first breeding success was as early as 1883 in France, after the parents had been acquired in 1872. California followed in the nineteenthirties.

In more recent times it appears that Sun Conures were only imported into Europe again in 1971, and the first breeding success was in Britain in 1973. Jim Hayward then imported a shipment himself, some of which he kept, but a number of which he sold. He kept his own birds in a colony for a number of years but failed to get any young in this situation, whereas other keepers did have success with birds from the same shipment. Later when he split his colony up and housed the birds in pairs, they soon laid eggs and reared young. They bred at various times of the year. A Dutch breeding award was given in 1978, and Dutch-bred birds were first exhibited at the Parrot Society show of 1980.

This Aratinga has never been available in large numbers.

Compared to the Jendaya the plumage of the Sun Conure can sometimes be rather untidy; they ruffle their feathers up which detracts somewhat from their appearance. However, it has nothing to do with illness, the only drawback being the derogatory effect on the look of the bird. A second, perhaps more serious, quality is their loud voice. However, they generally only make use of it when disturbed by unexpected occurrences and at sunrise and nightfall. Otherwise they are easy-to-keep and hardy birds, which can be seen to their best advantage in spacious aviaries, as their superb colouring is best displayed during flight.

The fact that a bird does not need all its claws in order to reproduce is proved by the experience of a German breeder: his hen had only one remaining claw and still reared young with success!

At one time I had a single cock, and as South American parrots generally like company I put it in with a single *Pyrrhura frontalis*, so that they could entertain each other. And they certainly did just that, for imagine my surprise when some time later I found eggs in the box, one of which was fertile! When something like this happens by accident you cannot help but be inquisitive as to the result, and I allowed the *Pyrrhura* hen to carry on incubating. The egg hatched and the chick was reared without any problems. It looked very similar to the hen; it was only a little larger and had yellowish-green colouring on the belly. Otherwise it was mainly green, and two years later the plumage was unchanged. A bird such as this must of course not be used for breeding purposes; it is to no advantage. Only pure-bred birds should be bred from.

Mutations

None known.

WEDDELL'S CONURE - *Aratinga weddellii*

Sub-species

None.

Origin of Name

Aratinga: shining parakeet.
Weddellii: Dr Hugh Algernon Weddell was a naturalist with a large French expedition to South America in 1843 to 1847.
Dutch: Weddell-aratinga, bruinkopparkiet.
German: Braunkopfsittich.
The name can be misleading as the head appears more blue than dusky (or brown in Holland and Germany). If the bird is studied carefully it can be seen that the head feathers are in fact a greyish-brown, but that they have blue edging. This edging gives the whole a blue impression.

Parents and Young

The head is greyish-blue, and the large eye ring is white. The shade of blue varies with the amount of light, and appears deeper in the summer months. Otherwise the bird is green, tending to yellow-green on the belly. The ends of the flight and tail feathers are blue, as in many Aratingas. Apart from a possible difference in the size of the head and bill it is impossible to distinguish between cocks and hens.
On leaving the box young birds are generally duller and their bellies are still fairly green. As usual the iris is dark to begin with; the naked eye ring is smaller in size.
As I have no personnal experience of aviary-bred birds, although they must have occurred, I cannot state with certainty when this species is sexually mature. However, it will most likely be at an age of two years.

Sizes and Weights

Length: 28cm (11in).
Weight: approximately 115g.
Ring size: 6mm.

22. The Weddell's Conure has a striking white eye ring.

Habitat and Habits

In the wild they prefer three habitats: savanna with scattered trees; trees along the rivers in the Amazon region; and (in Colombia) woodlands in the tropical zone on the east side of the Andes mountains. The Weddell's Conure is a true lowland bird, which lives in small flocks of up to about 20. In the greater part of its distribution range it is fairly to very common. Nevertheless, not much is known about it. The size of the population is stable.
These Aratingas are not very noisy. Usually they only use their voice when disturbed, and for a short time in the morning and evening.

Diet

Wild: information about this is scarce. However, these species probably has similar feeding habits to most other Aratingas: seeds, berries, nuts, leaves, buds, etc.
Aviary:
1. sunflower seeds, apples, grapes, oranges, soaked raisins, millet, canary seed, hemp, peanuts and bread soaked in milk. In addition occasionally maize, carrot, beetroot and banana. This keeper's birds hardly touch greenfood, but they like willow twigs;
2. a standard parakeet seed mix with fresh fruit such as apples, blackberries, bananas and grapes, depending on availability. Also now and then rowan, hawthorn and elder berries, and the heads of dandelions. In the breeding season white bread soaked in milk;
3. an enriched seed mix for large parakeets, germinated seeds, lots and varied fruits and nectar;
4. a seed mix for large parakeets, rearing mix, pieces of apple, strawberries and cherries.

Nesting Sites

A British keeper discovered that his three pairs preferred boxes which hung crooked above those hung straight; the chosen boxes also all had a tunnel entrance 15cm (6in) long. Unfortunately, he does not given any other dimensions. Other successes have been achieved in boxes measuring: 47x15x15cm (18.5x6x6in); 50x30x30cm (19.5x12x12in); 60x40x40cm (23.5x16x16in); 30x24x24cm (12x9x9in), with a 7cm (3in) entrance hole; 75x25x25cm (30x10x10in); and 45x28x28cm (18x11x11in).

Breeding Process

Birds here can start breeding fairly early in spring, and many keepers even mention the period of mid January to the beginning of February. Two to four eggs are laid which are then incubated for on average 24 days. The eyes open after about ten days. After about 18 days the hen ventures out of the box for longer periods. The young leave the nest after 50 to 60 days.

If pieces of half rotten wood are placed in the box, the hen will spend some time chewing it up prior to laying the eggs. This is a good way of stimulating the breeding instinct.

The British aviculturist mentioned in 'Nesting sites' had three eggs in January 1986 from a pair which had been bred by him in 1983. However, the chicks died after a week, and another clutch of again three eggs was laid in April. All went well this time. A second pair laid in June.

Another British keeper's birds laid three eggs at the end of May and the beginning of June, which hatched after about 24 days. The eyes opened after 12 days, and the young left the box after 63 and 61 days.

The first breeding success in the Netherlands was probably in 1983, when Dekker of the Hague had three eggs, all of which hatched. The chicks were ringed on the twelfth day, and they left the box after about two months. A second clutch followed in May of the same year; the first young had then already been separated from their parents.

It is worth mentioning that a number of breeders report that their birds only eat egg mix when they have chicks.

General Remarks

This Aratinga first reached London in 1923. It was unheard of for some time after that, until it was suddenly once again imported in fairly large numbers in the mid-nineteenseventies. The first breeding success was achieved in Germany in 1976; America followed in 1978. In the Netherlands a breeding award was given in 1983.

I saw my first Weddell's Conures at an importer's on 8th August 1979, and I bought four, which I housed together in one aviary. Some time later one had died. On 8th April 1982 I found an egg in the box. Unfortunately I found the hen dead on the nest four days later. It transpired that she had been unable to lay a soft-shelled egg. Out of necessity I then put the first egg under a breeding *Pyrrhura frontalis* hen, but it proved to be infertile. I had two remaining birds. One laid that same year on 23rd and 25th April, but on the 26th the hen was looking poorly and she died the next day. Once again this was the result of a soft-shelled egg, despite the fact that they had received sufficient grit. The two eggs were again placed under the same *P. frontalis* hen. One of them hatched but the chick died as it was much smaller than the young Pyrrhuras. What with one thing and another my experience of Weddell's Conures is unsuccessful. The remaining cock was finally sold.

Like most other Aratingas these are hardy birds, which can normally speaking spend the winter in an unheated night shelter.

Mutations

A birdkeeper in Rio de Janeiro is in possession of a yellow mutation.

23. *Aztec Conures are among the rarest aviary birds.*

JAMAICAN CONURE - *Aratinga nana nana*

Sub-species

1. *Aratinga nana nana*   - Jamaican Conure
2. *Aratinga nana astec*   - Aztec Conure
3. *Aratinga nana vicinalis*   - Eastern Aztec Conure

Number 2 is smaller than number 1, but has relatively longer wings; the green tends to yellow and is paler, and the brown on throat and breast is also paler. Finally the bill is slightly smaller.

The beak of number 1 is reported to be clearly lighter in colour than that of the other sub-species.

Number 3 is very similar to number 2, but its plumage is brighter green; throat, breast and belly are much greener with less brown.

A fourth sub-species is sometimes distinguished, the *Aratinga nana melloni* (Honduran Aztec Conure), which looks like the other Aztecs, but with a more olive coloured head, and a paler green neck.

Aztec Conures are recognized by some systematists as a separate species. However, they are very similar to the Jamaican Conure. There is also some contention as to which the nominate form is, the *Aratinga nana nana* with the *Aratinga nana astec* as sub-species, or vice versa. Here, as elsewhere in this book, I follow Forshaw. Such a discussion is not very important to the average birdkeeper.

Origin of Name

Aratinga: shining parakeet.
Nana: dwarf.
Astec: the Aztecs were the original inhabitants of Mexico.
Vicinalis: similar to the neighbours.
Dutch: Jamaicaparkiet.
German: Jamaikasittich.

Parents and Young

At first sight these mainly green and brown coloured birds appear rather dull. However, if you look more closely you will see that they are nevertheless unusual. The colouring on the belly, which is not very well visible in the photo, is particularly beautiful. A splendid pattern of darker vertical stripes which can be seen against an olive brown background, gives this species a unique distinguishing feature. It also has a few orangeish-red feathers round the nostrils.

Young birds are almost identical to their parents; the colours are slightly duller, and like all young birds they have dark irises.

Sizes and Weights

Length: Jamaican 26cm (10in), the others 24cm (9.5in).
Weight: 85g.
Ring size: 6mm (the leg size is 4.5mm, which means that a 5mm ring is rather on the tight side).

Habitat and Habits

The nominate form is common throughout Jamaica, particularly in the lowlands, wooded hills and on the lower mountain slopes. It is less at home higher in the mountains, and is most numerous in the wet forests. The Jamaican Conure tends to remain in the lower and middle branches of the trees, and never flies over the crowns.
The Aztec Conure lives in the fringes of deciduous woods mainly in the wet and semi-arid areas along the Caribbean coast up to an altitude of 750 metres.
The Eastern Aztec Conure is a typical lowland bird, and it never enters deep into woods.
Jamaican Conures are usually found in pairs, or small groups of between five and thirty birds. When certain sorts of fruit are ripe they form larger flocks of as many as 800 birds, and also move into the drier regions.
The pair bond is maintained within the group, and this is particularly noticeable when they settle in trees; each pair sits slightly apart and the two birds preen each other.
It has been reported that these Aratingas are not aggressive, and that they can easily be housed with other species. However, you should bear in mind that experience with this species is minimal, and it is therefore advisable to go about this with caution.
The species is fairly to very common in the wild and the population is stable.

Diet

Wild: they feed on berries, blossoms, fruits, seeds of grasses and weeds, and probably greenery; most of this is found in trees and bushes.
Aviary: all sorts of seeds, greenfood, fruit, nuts, vitamins (in the drinking water), soaked seeds of grasses and weeds, and soaked millet.

Nesting Sites

Jamaican and Aztec Conures are known to nest in cavities in termite hills, which they excavate in order to lay a maximum of five eggs. They may possibly also use holes in trees. Nests of Aztec Conures have more than once been found in cracks and hollows in limestone rocks.
Nothing is known about the breeding habits of the other sub-species.
Nest-boxes in aviaries probably do not need to be very large. One measuring 35x15x15cm (14x6x6in) should be sufficient; in termite hills the birds will also not have much room available. Considering their nesting habits it is perhaps worth trying a box with a tunnel entrance.

Breeding Process

Apart from the information given above there is very little to be found about the breeding process. This is mainly due to the scarcity of the bird and the correspondingly small number of breeding results. The reports which do exist are vague.
Although a breeding award was given in the Netherlands in 1988, at the time of writing no report of it has been published.

General Remarks

The Aztec Conure was first bred successfully in Jamaica in 1903.
The nominate form probably first arrived in Europe in 1981 (Germany).
Very few birdkeepers have the pleasure of being in possession of these birds as the species is extremely rare in avicultural collections. At the time of writing they are occasionally being imported into Germany.

Mutations

None known.

PETZ'S CONURE - *Aratinga canicularis canicularis*

Sub-species

1. *Aratinga canicularis canicularis*   - Petz's Conure
2. *Aratinga canicularis eburnirostrum*   - Western Mexican Petz's Conure
3. *Aratinga canicularis clarae*   - Petz's Conure

These three appear almost identical. An important difference is the colour of the bill. The upper mandible of all three is horn-coloured. The lower mandible of number 1 is also horn-coloured, but that of number 2 is greyish-brown, and of number 3 blackish. Moreover, the sides of the belly of number 2 are greener than the yellowish-green of number 1, and the orange forehead stripe is narrower. The stripe extends as far as the naked eye ring in the first two sub-species.

Number 3 only has a small orange spot which does not reach the eye ring, and in between is a blue stripe, which runs from the crown to the bill. There is also some difference in the colours of the throat and belly: those of number 1 are pale olive, and those of both the others are greener.

Origin of Name

Aratinga: shining parakeet.
Canicularis: belonging to the Dog Star (Sirius), which rises in the dog days between 23 July and 23 August.
Eburnirostrum: ivory beak.
Clarae: Clara was the first name of the wife of the American discoverer of this sub-species, Chester C. Lamb (1883-1965).
Dutch: Petzparkiet.
German: Elfenbeinsittich
USA: Halfmoon Conure.

Parents and Young

Both sexes have the same plumage. They are principally green, with various shades ranging from olive to lighter green on the belly, depending on the sub-species. The crown has a blue sheen.
They also possess a striking large naked yellowish-white eye ring. It is sometimes claimed that the hens have less orange on their foreheads.
As with all young birds the irises are brown; the orange marking on the forehead is smaller, and the plumage is generally duller.

24. The orange of the forehead of this Western Mexican Petz's Conure extends to the eye ring.

Sizes and Weights

Length: 24cm (9.5in) for all sub-species.
Weight: cock 75g, hen 70g.
Ring size: 5mm (leg size 4.2mm).

Habitat and Habits

The distribution range of the Petz's Conure extends along a large proportion of the west coast of Central America, from the lowlands and into the mountain slopes up to about 1500 metres. It prefers deciduous woods and dry scrub, but is also found in more open savanna country with scattered trees. Number 3 prefers to inhabit the wetter areas, such as marshland and land close to water courses. Outside the breeding season they travel about in flocks of up to two hundred birds in search of food. Every morning they leave their roosting-sites in the fringes of woods or along water courses in order to fill their empty stomachs. They rest during the hot hours in the afternoon, and return to their roosts in the evening after they have eaten enough to last them through the night. This pattern repeats itself year after year.
This species is very common and has adapted itself very successfully to the transformation of large parts of the region into semi-open cultivated grassland.

Diet

Wild: various seeds, nuts, berries, and fruits of mango and coconut trees, various sorts of fig, and insects and their larvae.
Aviary:
1. a seed mix for large parakeets, seeds of various grasses, apples, bananas, soaked maize, raisins, and bread soaked in milk or water;
2. sunflower seeds, mung beans and oats in germinated form, pigeon feed which has been soaked for 24 hours, and a mixture of pigeon feed and mung beans which have been boiled for two minutes. Egg mix is added to the last. Also boiled maize, pieces of fruit, berries and greenstuff, a mixture of white and striped sunflower seeds and millet. A multivitamin and mineral preparation, and calcium is sprinkled over the feed. In addition a few shrimps are also occasionally given.

Nesting Sites

Petz's Conures nest mainly in cavities, which they excavate, in the nests of only one species of tree termite. This is true for all three sub-species. It is reported that in some areas half of all the termite nests are used by Petz's Conures, usually only one pair in each. It takes them about a week to dig a tunnel 7cm (3in) in diameter and 30cm (12in) long, with at the end a nest cavity measuring about 15x20cm (6x8in). It seems that the hen does most of the work. Occasionally they choose the deserted nest of a woodpecker in a hollow trunk or branch.
Despite the fact that the nesting site in the wild is somewhat unusual, standard nest-boxes

are usually supplied in aviaries. As the breeding results are far from good, there is something to be said for trying to imitate the situation in the wild. In Walsrode Bird Park in Germany success has been recorded with a 60cm (24in) tall and 30cm (12in) square box, the 8cm (3in) entrance hole of which was 30cm (12in) above the floor.

Breeding Process

Little is known about this, and I have not come across many good breeding reports. Between three and five eggs are laid. The only successes mentioned were from the Walsrode Bird Park. A pair of the Western Mexican sub-species were bred in the parrot house in 1988, where they were housed with a pair of Inca Cockatoos in an aviary 2.20m (7ft) long, 3m (10ft) wide, and 2.10m (7ft) high. Three eggs were laid at two-day intervals. Incubation began after the first egg, and the incubation time was 23 days. The chicks left the box after 49 days. They were almost identical to their parents; the orange forehead stripe was narrower and the irises were darker. They remained with their parents for more than six months without any problems arising.

It is rather surprising that in this day and age, when aviculturists have been extremely successful in breeding the most rare species, so little information is available about such a relatively easy bird as the Petz's. Is it that breeders do not find successes worth reporting? This would be the wrong attitude given that the most experienced birdkeeper continues learning all their lives and can still occasionally, be confronted by an unexpected situation. Therefore it is always worth reporting success on each species. My comments on these birds in the next section make it clear that it should be possible to breed them with success.

General Remarks

The first members of this species to reach Europe appeared in London Zoo in 1869. Reports of the first breeding success came from California in 1929. Germany followed in 1932, then Czechoslovakia (Prague Zoo) in 1971, Britain in 1976 and Italy in 1977. An award was given in the Netherlands in 1980.

The sub-species numbers 2 and particularly 3 are virtually always available. Number 1 is rarely found.

Unfortunately, the Petz's Conure has never been popular in Europe, despite the fact that it is very suitable as an aviary bird. It is not too big, does not require spacious accomodation, is usually not noisy and is not too destructive. With a little time and effort it should be possible to achieve more breeding successes. However, in America Petz's Conures are extremely popular. Many people keep them in cages as indoor pets and it appears that they are well suited to this kind of existence. They often become extremely tame, and some even learn to speak a few words.

The Petz's Conure is sometimes confused with the Golden-crowned Conure (*Aratinga aurea*), which is described later. However, there are clear differences between the two and any birdkeeper who has seen both and compared them will never confuse them again. Although there are more differences, it is sufficient to look at the bills. The Golden-crowned's is deep black, whereas the upper mandible of the Petz's is horn-coloured. A comparison of the two photos shows the other points to look out for; whether the orange

25. The orange mark on the forehead of this Petz's Conure (A. c. clarae) does not extend as far as the eye ring.

on the forehead extends to the eye or not, and whether the eye ring is naked or feathered. Finally, the Golden-crowned is slightly larger: 26cm (10in) as opposed to 24cm (9.5in).

Mutations

There is known to be a yellow (or dilute?) mutation of number 2 in Germany, whose belly is bright yellow instead of green. The neck and breast are greenish-yellow, and the cheeks and ear coverts are orangeish-brown. The wing feathers have yellowish edging, and the lower mandible is horn-coloured. The orange marking on the forehead remains. So far the bird has not been used for breeding.

26. *This St Thomas Conure can be easily recognized by the amount of yellow on the head. The Bonaire Brown-throated Conure has more yellow on the crown.*

ST THOMAS CONURE - *Aratinga pertinax pertinax*

Sub-species

1. *Aratinga pertinax pertinax*   - St Thomas Conure
2. *Aratinga pertinax xanthogenia* - Bonaire Brown-throated Conure
3. *Aratinga pertinax arubensis*   - Aruba Brown-Throated Conure
4. *Aratinga pertinax aeruginusa* - Brown-Throated Conure
5. *Aratinga pertinax griseipecta* - Colombia Brown-throated Conure
6. *Aratinga pertinax lehmanni*   - Lehmann's Brown-throated Conure
7. *Aratinga pertinax tortugensis* - Tortuga Brown-throated Conure
8. *Aratinga pertinax margaritensis* - Margarita Brown-throated Conure
9. *Aratinga pertinax venezuelae* - Venezuela Brown-throated Conure
10. *Aratinga pertinax chrysophrys* - Guiana Brown-throated Conure
11. *Aratinga pertinax surinama*   - Surinam Brown-throated Conure
12. *Aratinga pertinax chrysogenys* - Golden-cheeked Brown-throated Conure
13. *Aratinga pertinax paraensis*   - Para Brown-throated Conure
14. *Aratinga pertinax ocularis*   - Brown-eared Conure

The differences between these sub-species can mainly be found by studying the varying amounts of brown and yellow on the cheeks and around the eyes. The rest of the plumage is almost identical in all cases, and it is sometimes extremely difficult to tell birds apart. Another complication is that there are hybrids between certain sub-species, which make things even more difficult. The distribution map shows clearly that the ranges of certain sub-species border on each other. Sometimes I cannot help wondering if it is correct to distinguish so many sub-species. However, as this classification seems to be generally accepted I will adhere to it here, and attempt to give a short summary of the most important differences. In order to do this I have used the information given by Forshaw and Arndt, as there are no other systematic summaries to be found. There is also probably no one else who has seen all fourteen sub-species. This means that although I give as much information as possible, I cannot personally guarantee its reliability. The illustrations give certain amount of additional assistance.

A. p. pertinax: can be recognized by the bright yellow cheeks, and narrow yellow forehead stripe.

A. p. xanthogenia: as the nominate, but the yellow on the head extends backwards over the crown to above the eyes. It is the most beautifully coloured sub-species.

A. p. arubensis: the orangeish-yellow is limited to a ring round the eye; the amount under the eye is greater than the amount above. The forehead is pale yellow; the dull greenish-blue on the crown runs almost right over the back of the head; lores, cheeks and sides of head are a mixture of light-brown and very pale orangeish-yellow; the ear coverts are yellow with brown edges; throat and upper breast are yellowish-brown.

A. p. aeruginosa: as *arubensis*, but with very little pale yellowish-brown on the forehead; the orangeish-yellow is limited to a narrow ring roung the eyes; throat, breast and sides of head are slightly darker and browner; the greenish-blue on the crown extends down into the neck.

A. p. griseipecta: as *aeruginosa*, but cheeks, throat and upper breast are olive grey, merging into green on the lower breast; no yellow stripes on the ear coverts; very little blue on the green crown.

A. p. lehmanni: as *aeruginosa*, but with a much wider orangeish-yellow ring round the eye; the greenish-blue of the crown is only at the front and does not extend into the neck; a very faint blue tinge at the end of the middle tail feathers; a slightly darker throat than *chrysophrys*; the outer barbs of the primary wing feathers are more green than blue.

A. p. tortugensis: as *aeruginosa*, but with more orangeish-yellow on the sides of the head, becoming paler on the lower parts of the sides of head and the throat; under wing coverts tending to yellowish-green; slightly larger.

A. p. margaritensis: the forehead is whiteish; the front of the crown is greenish-blue; lores, cheeks and ear coverts are olive brown; orangeish-yellow round the eye, and more extensive under it; throat and upper breast are pale olive.

A. p. venezuelae: as *margaritensis*, but generally paler and more yellowish; the inner barbs of the tail feathers have yellow edges; less orange on the lower belly than *chrysophrys*.

A. p. chrysophrys: as *margaritensis*, but throat and cheeks are a darker, purer brown; the forehead is pale yellowish-brown.

A. p. surinama: as *chrysophrys*, but the orangeish-yellow round the eye also covers the lores and extends to the lower mandible; narrow orangeish-yellow forehead stripe; throat and breast are pale yellowish-green instead of brownish.

A. p. chrysogenys: the plumage is generally darker than the other sub-species; lacks pale forehead stripe; crown is dark blueish-green; outer barbs of flight feathers mainly dark blue; throat, breast and sides of head are dark brown; lower belly has dark orange sheen.

A. p. paraensis: forehead and crown are blueish-green; feathers on back of neck are dark green, with light brownish-yellow edges; dark-green back; outer barbs of flight feathers are dark-blue; throat, breast and sides of head are brown, darker on ear coverts and above the eyes; orangeish-yellow round the eyes; lower belly is orangeish-yellow with brown, merging into green above; underside of tail contains yellowish-green; iris is red.

A. p. ocularis: entire forehead and crown are green, sometimes with a blueish sheen; orangeish-yellow in front of and, to a greater extent, under the eyes; lores and sides of head yellowish-brown; throat and breast are brownish, paler than the sides of the head; the belly is yellowish-green, becoming more yellow further down.

While reading this it will undoubtably be difficult to form a picture of each separate sub-species. In order to make life easier a number of extra tips now follow.

Three of the fourteen can give no cause for confusion: *pertinax*, *xanthogenia* and *surinama*. The first two are the only ones with entirely yellow cheeks, but the crown of *xanthogenia* is also yellow. Of the others *surinama* is the only one with yellow lores; the area between eye and beak.

Other clues can be found by studying the eye ring, forehead and the top of the head above the eyes, the crown. *Aeruginosa*, *griseipecta* and *tortugensis* have only the odd yellow feather round the eye; *lehmanni* and *paraensis* have a closed yellow circle, and *arubensis*, *margaritensis*, *venezuelae*, *chrysophrys*, *chrysogenys* and *ocularis* have more yellow under the eye.

Arubensis has a pale yellow forehead; *aeruginosa*, *griseipecta*, *lehmanni*, *tortugensis* and *chrysophrys* have pale yellowish-brown; *margaritensis* and *venezuelae* have off-white; *chrysogenys* dark brown; *paraensis* blueish-green; and *ocularis* green.

The crown of *arubensis*, *aeruginosa* and *paraensis* is blueish-green; of *griseipecta*,

27. *One of the sub-species of the St Thomas Conure, the Aruba Brown-throated Conure.*

28. *The small amount of yellow round the eye indicates that this is a Brown-throated Conure.*

tortugensis and *ocularis* green; of *chrysogenys* blue; the crown of *lehmanni, margaritensis, venezuelae* and *chrysophrys* is blueish-green at the front and green at the back.

Origin of Name

Aratinga: shining parakeet.
Pertinax: with a strong grip, tenacious.
Xanthogenia: with yellow cheeks.
Arubensis: from the island Aruba.
Aeruginosa: covered with copper rust.
Griseipecta: with grey breast.
Lehmanni: after W. Lehmann (1882-1968), a German poet and writer with a good knowledge of natural history.
Tortugensis: from the island Tortuga off the coast of Venezuela.
Margaritensis: from the island Margarita off the coast of Venezuela.
Venezuelae: from Venezuela.
Chrysophrys: with golden eyebrows.
Surinama: from Surinam.
Chrysogenys: with golden cheeks.
Paraensis: from the Brazilian province of Para.
Ocularis: relating to the eye.
St Thomas: a small island in the Caribbean.
Dutch: St. Thomasparkiet.
German: St. Thomassittich; the sub-species are called Braunwangensittiche.

Parents and Young

It is not possible to determine sex visually. There are therefore two alternatives: you can either purchase a number of birds and house them together until they have formed pairs, or you can let them undergo an endoscopic examination.
If you wish to rear chicks you must make sure that the parent birds have exactly the same colouring; if there is any difference between the two there is a good chance that they will be two different sub-species.
Young birds of the first two sub-species have much less yellow on the head than adults. They also have brown cheeks, the upper mandible is horn-coloured, and the throat and breast are greener. The plumage of the young of all the other sub-species will appear generally duller, but particularly with regard to the brighter colours, and especially yellow, which is preceded by shades of green and brown.
It will be clear from the above that when trying to determine to which sub-species a bird belongs, it is important to take its age into account. You must know at least whether it has developed its adult plumage.

Sizes and Weights

Length: St Thomas Conure 25cm (10in), the other sub-species 24 to 26cm (9.5 to 10.5in).

29. *The least brightly-coloured member, the Colombia Brown-throated Conure.*

30. *Some Guiana Brown-throated Conures display more brown on the belly than the bird shown here.*

Weight: St Thomas cock 95g, hen 90g; the sub-species respectively 100 and 85g.
Ring size: 6mm.

Habitat and Habits

No fewer than eight sub-species inhabit areas throughout a large part of South America north of the Amazon, and five occur on a few of the Caribbean islands.
The St Thomas originally only occurred on Curaçao; it was probably taken to the island of St Thomas more than a century ago. The sub-species *A. p. xanthogenia* is found only on Bonaire. They are both common on their respective islands, and feel at home in almost all the available habitats.
The sub-species living on the continent are found in areas of savanna and scrub, and on agricultural land. They also occur in lesser numbers in the mangrove woods and in the coastal region. The only habitat they avoid completely is dense forest. They are principally lowland birds, although they are occasionally encountered in small numbers up to an altitude of 1200 metres.
This species is fairly to very common throughout a large part of its distribution range. It is very tolerant of ecological changes, and adapts quickly to new conditions. It is even found in villages and towns. As a result most of the sub-species is on the increase rather than the decrease.
These parrotlikes are usually seen in pairs or small groups; it is only found in larger flocks in places where there is an abundant supply of food.
The breeding pairs leave the group during the breeding season.

Diet

Wild: various seeds from grasses and weeds, cactus fruits, acacia seeds, buds, nectar, various cereals (they can cause quite a lot of damage to crops), and probably insects and their larvae.
Aviary:
1. the Venezuela Brown-throated Conure mentioned below ate only a seed mix for large parakeets, millet sprays and apples. It refused germinated seeds and egg mix;
2. parakeet seed mix containing lots of paddy (unmilled) rice, millet sprays, rusks, brown bread, egg mix, germinated rapeseed, privet, chickweed and golden rennet apples.

Nesting Sites

In the wild, nests are sometimes found in crevices and cracks in limestone rocks, in steep banks of earth or limestone and in holes in old trees, but the majority of birds nest in still inhabited nests of tree termites. It seems that some sub-species nest in the roofs of Indian huts. Nests in termite nests often have an entrance tunnel with a diameter of 7 to 10cm (3 to 4in); one such nest which was opened for examination was found to have a tunnel 50cm (20in) long with a larger cavity about 25cm (10in) in diameter at the end.
On the islands of the Dutch Antilles birds have been found breeding in small colonies:

four or five pairs in the trunk of a dead date palm, and a number in holes dug in a steep sandy bank.
If the conditions are favourable breeding can take place throughout the year.
In the aviary they are no more fussy than in the wild and once they are ready to breed they readily accept any reasonable box. It must of course not be too large; 35 to 40cm (14 to 16in) tall, with a floor 15cm (6in) square is sufficient. Entrance hole: 6cm (2.5in).

Breeding Process

A clutch usually numbers between four and seven eggs. The cock often spends the night with its mate on the nest. It is normal for the whole family to return to the box just before nightfall for a considerable time after the young have left it.
In the case of the Guiana Brown-throated Conures bred in Britain mentioned in the next section, two pairs were housed in an aviary which was nine metres (30ft) long. In 1954 the hens laid four and five eggs, but without result. A year later two chicks hatched in one of the nests on 12th and 13th July. The young left the box on 30th and 31th August, both after 49 days. They were identical to their parents. Rondhuis of the Netherlands reported his first success in 1979. His pair produced five eggs, laying every other day from 9th May. Three eggs hatched after 24 days, and the chicks were given 5mm rings.
The Venezuela Brown-throated Conure was bred in Switzerland in 1982. The breeding pair were housed in an aviary measuring one metre (3ft) square and 2 metres (6.5ft) high. An 80cm (31.5in) long hollowed-out branch with a diameter of 22cm (8.5in) and a 6cm (2.5in) entrance hole was provided. Three eggs were laid, which unfortunately proved to be infertile. A second clutch of four eggs, laid at three-day intervals, followed from 6th June. The first chick hatched on 3rd July, followed two days later by the second. The embryos of the other two had died at an early stage. About seven weeks later the two young left the nest within two days of each other. They were feeding with their parents from the container after only one day. After a month the parents began to peck them, and eventually the young had to be removed.
Since 1983 my Guiana Brown-throated Conures have had a number of successful breeding seasons. I came across two imported birds unexpectedly in 1982 and got results immediately the following year. On 4th June the first of four eggs was laid, all of which hatched from 29th June onwards. Assuming that the hen started sitting after the second egg this gives an incubation period of 23 days. The first young left the box on 17th August, on the 49th day. In subsequent years the number of eggs varied from four to six, and they occasionally bred twice in one year. As far as I can judge the incubation period was always 23 to 24 days, and the young left the box after between 47 and 51 days. On one occasion this ran to 57 days, but that was the last-born of a second clutch and it was badly fed.
I also have a pair of St Thomas Conures. For a number of years the hen occasionally laid one egg, but did not incubate. However, in June 1989 things improved. Three eggs were laid and the hen carried out her task admirably.

General Remarks

The first success with the St Thomas Conure is reported to have been in 1949 in the USA.

The first report from the Netherlands dates from 1968, when three young were reared from four eggs. They were ringed on the twelfth day.
The St Thomas should really be called the Curaçao Conure, as it originally comes from that island, and was introduced to St Thomas.
Wassenaar Zoo in the Netherlands acquired a pair of Bonaire Brown-throated Conures in 1979, and one young was reared that same year, followed by another four in 1980.
The Brown-throated Conure was first bred in Britain in 1908. The first Colombia and Tortuga Conures were probably bred in France in 1981 and 1980 respectively. The first young Venezuela Conures to be reared in Europe first saw the light of day in Switzerland in 1982.
The Guiana Conure is the sub-species which is most common in collections, and which has been bred most often in captivity. Britain was the first in 1955, and also with the Brown-eared in 1915.
In the Netherlands breeding awards have been given for a number of sub-species: St Thomas Conure in 1982, the Bonaire Conure in 1983, the Brown-throated in 1981, the Guiana in 1978, the Surinam in 1985. The first Dutch home-bred Surinam and Bonaire Conures were exhibited at the Dutch Parrot Society shows of 1980 and 1982 respectively.
Up until a few years ago Dutch people returning from the Dutch Antilles to the Netherlands occasionally brought their pet St Thomas Conures back with them. I have such a bird, which used to belong to the van Doorn family of IJsselstein. They adopted her (the bird has laid an egg in the cage a number of times) from an aquaintance and kept her as a pet during their stay in Aruba. Although the St Thomas Conure originally comes from Curaçao they saw many of them flying about freely in Aruba. Funnily enough this is not mentioned in any of the literature, which only says that they are to be found on St Thomas and Curaçao. This particular bird eventually started feather-plucking, and its owners wondered whether it might stop if they found a mate for it, and also whether they would breed. It has now been in one of my aviaries in the company of a cock since June 1989, bu still shows no interest in him.

Mutations

None known.

31. The Cactus Conure clearly distinguishes itself from the Brown-throated Conures with its light-coloured bill and its clear yellow belly.

CACTUS CONURE - *Aratinga cactorum cactorum*

Sub-species

1. *Aratinga cactorum cactorum*   - Cactus Conure
2. *Aratinga cactorum caixana*   - Pale Cactus Conure

The plumage of number 2 is slightly paler, and the throat and breast are deeper brown. However, the differences with number 1 are slight, and even with the two sub-species next to each other it is fairly difficult to tell which is which.

Origin of Name

Aratinga: shining parakeet.
Cactorum: of the cacti.
Caixana: the Caixa is a river in the Brazilian province of Bahia.
Dutch: cactuspakiet.
German: Kaktussittich.

Parents and Young

The sexes are almost identical in appearance. Hens may be very slightly paler in colour, but in practice it is impossible to determine the sex visually.
They have a striking naked eye ring. Their breasts are brown and their bellies are light yellow, and the division between the two is very sharp. This feature alone clearly distinguishes them from the Brown-throated Conures.
Some birdkeepers confuse the two, but the photos show that this is quite unnecessary. The colour of the Brown-throated's beaks ranges from dark to black, whereas the beak of the Cactus Conure is horn-coloured, and none of the Brown-throateds has a naked white eye ring. Nevertheless, some experts consider the Cactus Conure to be a sub-species of the St Thomas species. I have my doubts about this.
The plumage of young birds is paler, and the belly is more yellowish-green. The crown is green, and the iris is of course dark. Birds are probably mature at the age of two.

Sizes and Weights

Length: 25cm (10in).
Weight: approximately 90g.
Ring size: 6mm.

Habitat and Habits

Their distribution range corresponds to the inland "caatinga" region of eastern Brazil. "Caatinga" is a South American term denoting the type of countryside covered with low thorny bushes, succulents, cacti and stunted trees. They also occur locally in more open semi-desert regions with scattered deciduous trees.

In his book "Conservation of New World Parrots" Ridgely suggests changing the name to Caatinga Conure, not only because of the fact that its connection with the caatinga region is so striking, but also because he noted that in the wild the species has no particular preference for cacti or their seeds. However, the name Cactus Conure is now so generally accepted, it would be difficult to change.

The general impression is that this species is fairly numerous; they are usually seen in pairs or in small groups of up to eight birds. The inaccessibility of their habitat is probably the main reason that very little is known about their habits. Field workers report that the population is stable. My experience of these birds is that they dislike wet and cold weather. In autumn they ruffle their feathers and sit hunched up, which at first glance gives the impression that they are ill. However, this is soon over once they are placed in a drier and warmer place.

Diet

Wild: seeds, berries, fruits, nuts, and probably blossom. Although they do eat the fruits of cacti they certainly do not form a large part of the diet.

Aviary:
1. small sunflower seeds, canary seed, white millet, a little hemp, oats, buckwheat and niger seed. Also millet sprays, and every other day peanuts, fruit (apples, grapes, a variety of wild berries, cherries, pears, oranges, pieces of melon, figs, etc) and greenstuff (chickweed, lettuce, spinach, dandelion heads, shepherd's purse, sprouts, carrots), supplemented particularly during the breeding season with soaked seeds;
2. a seed mix for large parakeets, supplemented daily with germinated seeds: sunflower seeds, safflower, millet, pigeon and canary seed; and mixed into the seed apples, oranges, peas, maize, beetroot, carrot, spinach, vitamins, cod-liver oil and wheat germ oil. In addition fruit and greenfood which are in season.

Nesting Sites

Nothing is known about the breeding habits in South America. It is possible that they nest in cavities in giant cacti as well as in the trunks and larger branches of trees.

In Germany they have been bred in a hollowed-out branch 20cm (8in) in diameter and 50cm (20in) tall. Successes have also been recorded using hollowed-out birch and beech branches 35cm (14in) tall and 20cm (8in) in diameter.

I have bred them in a fairly small box 40cm (16in) tall, a floor of 15cm (6in) square, and an entrance hole 6cm (2.5in) in diameter.

Breeding Process

Very little more is known about this species' breeding habits in captivity than in the wild. They can become rather aggressive during the breeding season. In 1984 I was compelled by circumstances to house a number of species in one aviary temporarily as I was doing some rebuilding. My Cactus Conures had to share their accomodation with two Dusky Pionus Parrots. All went well until the former showed the first signs of preparing to breed. From that time onwards the Pionus were no longer safe, and the Conures chased them about continuously. In the end I had to separate the two species and from that moment on the Cactus Conures were much quieter.

They had eggs for the first time in 1983, but they were infertile. Unfortunately, the same was true of two clutches of four eggs which were laid in 1984. Things seemed to be going better in 1985. I let the birds into the outside aviary at the end of March, and from 18 June no less than six eggs were laid at intervals of two (and once three) days. Of these only two proved to be fertile and they hatched on 14th and 15th July. I do not know exactly when they left the box as I was on holiday at the time. Unfortunately, a short time later a cock Red-masked Conure which was housed in the neighbouring aviary managed to get into that of the Cactus Conures. This resulted in the death of the Cactus cock and one of the chicks. The pleasure had then worn off to such a degree that I eventually sold the remaining birds.

From the five eggs which had been laid, a German breeder received three chicks, all of which thrived. Their eyes opened on the twelfth day, and they were ringed on the fourteenth. The first feathers started to appear after three weeks. They left the box after about 42 days, and were somewhat helpless for the first twenty-four hours. Generally speaking South American Parrots are much more self-reliant immediately after leaving the nest than, for example, the Australian species. They bred successfully a second time in the same year, producing a clutch of four eggs. A year later they successfully bred twice again, when five young were reared from two clutches of five eggs.

Other breeders report that chicks remained in the box for 51 and 53 days after hatching.

General Remarks

The first breeding success with this species was in Paris in 1883. Britain followed some years later in 1914. The first report from Denmark dates from 1963, and from Germany 1973. In the Netherlands an award was first given in 1987, although I had reared young in 1985. The first Dutch-bred bird was exhibited at the Parrot Society show of 1984.

This species only occurs in Brazil, because of the export ban no more of these birds will arrive in Europe. This means that we shall have to make do with the birds already here, and unfortunately their numbers are very small.

These birds hardly ever chew and they make very little noise. These characteristics, which are of course positive ones for birdkeepers, do not fit in with the general impression which many have of the Aratingas.

Mutations

The Brazilian Nelson Kawall is in possession of a yellow mutation of this Aratinga.

GOLDEN-CROWNED CONURE - *Aratinga aurea aurea*

Sub-species

1. *Aratinga aurea aurea*   - Golden-crowned Conure
2. *Aratinga aurea major*   - Greater Golden-crowned Conure

The plumage of both species is identical. The only difference is that, as the name suggests, number 2 is four centimetres larger. However, birds belonging to 1 also vary in size; as a rule those from Argentina are larger than those of the same species from the estuary of the Amazon, although they never reach the size of number 2.

It is doubted by some whether a distinction should be made between the two sub-species. Indeed there is a gradual transition between 1 and 2 with respect to the size of the birds, and it is a normal phenomenon for members of one species to become larger the closer to the equator they live.

Origin of Name

Aratinga: shining parakeet.
Aurea: golden.
Major: larger.
Dutch: goudvoorhoofdparkiet.
German: Goldstirnsittich.

Parents and Young

The orange-coloured patch on its forehead extends from the nostrils to the crown, although it does vary in size. Between this patch and the eyes is a blue stripe which completely edges the peach over the crown, and also extends to below the eyes. Most birds have a ring of peach orange feathers round the eyes, but some have an almost naked white eye ring with the odd orangeish feather.

As with most Aratingas there are no fully reliable differences between the sexes. The shape of the head and bill might give an indication, and it has been reported that the bellies of hens are greener than the cocks' yellowish-green, but it is nevertheless very risky to try and choose a pair using these methods.

The same is true for the amount of feathers round the eye; the hen of one particular breeding pair had far more than the cock. Young birds look very similar to their parents, but have less orange and blue on their heads, and their bills are noticeably paler. The feathers in the eye ring may also not be present. Their irises are dark. The young of a British birdkeeper only developed the peach colour in the eye ring after nearly two years.

32. At first sight the Golden-crowned Conure looks similar tot the Petz's Conure. However, the black bill removes any possible doubt.

Sizes and Weights

Length: 26cm (10in).
Weight: approximately 105g.
Ring size: 6mm.

Habitat and habits

The only other species to have a larger distribution range is the White-eyed Conure. The Golden-crowned Conure is found in open and semi open savanna. In the Amazon basin it occurs only in the isolated and scattered "campinas": these are areas of sand on which bushy vegetation grows, and which are often surrounded by damp forests on more fertile ground. Towards the south the numbers of Golden-crowned Conures increases and they become more evenly distributed as a result of the drier more open woods found there. It occurs throughout this region and is commonly seen. With the disappearance of the woods it may even be increasing its territory and numbers, particularly as it is quite adaptable and can even be found close to human habitation. This species is not very shy and takes little notice of its surroundings whilst foraging.
They are usually found in pairs or in groups of between ten and thirty birds, occasionally more. They spend a great deal of the day searching for seeds and fruits in trees and bushes or on the ground. Golden-crowneds are strong fliers which appreciate a long aviary. They can also be very noisy; they generally only use their piercing voice if they are excited or if something frightens them. The longer they spend in one aviary and the more they grow accustomed to their surroundings the less noise they make; they are then only heard in the morning and evening.

Diet

Wild: mainly seeds, fruits, berries, nuts, and insects and their larvae. These components have been found in the stomachs and crops of examined birds.
Aviary:
1. a standard seed mix, carrots, apples and spinach, with once a week a multivitamin preparation in the drinking water. Bread soaked in milk was refused, but cuttlefish was eaten;
2. mixed parakeet seed, apples, rosehips, germinated seeds, soaked bread and egg mix. Greenfood is almost completely refused;
3. a good seed mix, a mixture of grated carrot, hard-boiled egg, vitamins and egg mix; in addition maize, rosehips, germinated seeds and willow twigs;
4. sunflower seeds, a seed mix for budgerigars, safflower seeds, bread soaked in milk, oranges, apples, berries, raspberries, maize, various seeding grasses and twigs fom fruit trees.

Nesting Sites

Although this species is fairly common little is known about its nesting habits. They nest

in the hollow branches or trunks of trees; a clutch usually consists of two to four eggs. In aviaries birds have been bred in a box 25cm (10in) tall with a 23cm (9in) square bottom, and in one measuring 50x25x25cm (20x10x10in) with an 6cm (2.5in) diameter entrance hole.

In another case birds were given the choice of three boxes measuring 90x30x30cm (35.5x12x12in), 60x25x25cm (23.5x10x10in) and 30x22x22cm (12x8.5x8.5in). They chose the smallest of the three.

Breeding Process

A pair in Britain began to feed each other two years after they had been imported. In the third year four eggs were laid and two young were reared. Their eyes opened after about three weeks. The originally white down was gradually replaced by grey. When they left the box after 55 days their plumage was not complete, which is unusual for Aratingas. Their bills were not yet completely black, and only a little of the orange colouring was visible on their foreheads. Behind the orange was a clear yellow spot.

In another case a pair produced six infertile eggs in February, followed by a similar clutch of six. Success came in May with a third clutch of five eggs. The first-born chick left the nest after 48 days.

A pair in the Netherlands produced the first egg on 28th April, which was followed by two more. The cock spent the night in the box with the hen. All the eggs hatched and the chicks were ringed after fourteen days. The three fully-grown young left the box after 55 days. They were as large as their parents and had almost the same colouring; they were slightly paler and lacked the coloured eye rings.

In a number of other reported cases only three eggs were laid, which suggests that the size of the clutch varies considerably. However, it is also possible that these smaller clutches were due to diet.

General Remarks

Golden-crowned Conures have been kept in captivity for more than a century. The first success was as early as 1880 in Dantzig where the birds bred in an aviary measuring only 1.5x1.4x0.5 metres. The first British success was achieved in 1926. Two eggs were laid in April, and according to the report it was only in August that the one young left the box. This seems rather on the long side. Denmark and Germany followed in 1932, and then America also in the thirties. Later successes were achieved in: South Africa 1958, Sweden 1959, Australia 1963 and Hong Kong 1967. In the Netherlands a breeding award was given in 1980.

As is the case with the Petz's Conure, the number of known breeding successes is small. However, this is probably due to the fact that the main reason for keeping these birds has not been in order to breed them. The attitude has been one of putting two birds together and waiting to see what happens. Fortunately, during the last few years keepers have been concentrating more on breeding.

A birdkeeper in Britain let these birds fly about freely with success. They bred in a hollow tree and some time later were flying about with five young birds. However, after this first success the hen disappeared. The cock then paired with a Nanday hen, and they produced two young. They looked more like their mother than their father, with respect to both size and colouring, but they did have the orange-coloured forehead. The black on the head was not so deep and less extensive. The man also kept African Greys, Alexandrines, Ringnecks, Plum-headeds and Moustache Parrakeets in the same fashion; however, birds of the Ringneck family tended to fly off. All the birds were given extra food in an aviary, although not all of them made use of it. It should be said that it is now illegal to allow birds to fly freely in Britain.

The Golden-crowned Conure is sometimes confused with the Petz's. However, there are differences which are obvious enough to cut out any chance of a mistake: it is sufficient to mention only the deep black beak of the Golden-crowned; that of the Petz's is mainly horn-coloured.

They tend to quarrel with their neighbours occasionally, particularly if they happen to be other South American Parrots, and even more so if they have eggs or young. In this case double wire mesh is not out of place, though it is better still to place a species next to them which does not react to their aggressive behaviour.

Outside the breeding season it seems that some keepers house them together with other species, however, unless they can be placed in a very large aviary, this does not seem a good idea to me.

Although they are not exactly rare as aviariy birds, you certainly will not come across them very often.

Mutations

A birdkeeper in Brazil has one blue mutation and two cinnamon birds. Unfortunately, no other information is available.

ACKNOWLEDGMENTS

My grateful thanks are due to the following who so readily complied with my requests for information, names, photographs, maps and translation.

I am particularly grateful to Cees Scholtz for traveling so many kilometers with me and for his endless patience in producing photographs of such a high standard (pages 12, 22, 32, 44, 50, 56, 59, 60, 66, 72, 75, 76, 80, 82, 96, 100, 102, 108, 114, 118, 124, 128, 130, 134, 136, 140 and 146). I wish to acknowledge my thanks to Horst Müller (Germany) for his permission to publish the photographs on the pages 69 and 88, and Sebastian Fuss (Germany) for the picture on page 92.

I should like to thank the many aviculturists who provided information at my request, and to those who gave permission to have their birds photographed by Cees Scholtz:

Siem van 't Hart (Holland)	: page 114
Theo van den Heuvel (Holland)	: pages 108, 146
Cees de Koning (Holland)	: pages 44, 50, 102
Piet de Koning (Holland)	: page 130
Willem Plomp (Holland)	: pages 82, 134 upper
Hans and Erika Prinz (Germany)	: pages 56, 72, 75, 118, 124, 128
Rudolph Prinz (Germany)	: page 59

The photographs on pages 60, 66, 76, 80, 96, 134 lower, 136 upper, 136 lower, 140 are the author's birds.

I am indebted to Rosemary Low for providing the correct English names, and to Bauke Vliegendehond for drawing the maps.

And last but not least, I am particularly grateful to Ian Borwell for his fine translation.

LITERATURE

Th. Arndt: "Südamerikanische Sittiche; Keilschwanzsittiche i.e.S." Walsrode 1981.

"Conservation of New World Parrots". Proceedings of the ICBP Parrot Working Group Meeting. Roger F. Pasquier, editor. St Lucia 1980.

J.M. Forshaw: "Parrots of the World". Willoughby 1989.

W. de Grahl: "Atlas Papageien und Sittiche der Welt". Band I Sittiche. Walsrode 1982.

R. Harris: "Breeding Conures". Redhill 1983.

H. Kremer: "Australian Parakeets and their Mutations". Noordbergum, 1992.

R. Low: "Parrots, their care and breeding". Poole 1980.

R. Low: "Endangered Parrots". Poole 1984.

R. Low: "The parrots of South America". London 1972.

H. Strunden: "Die Namen der Papageien und Sittiche". Walsrode 1986.

Avicultural magazines 1980 - 1992:
    Australian Aviculture
    AZ-Nachrichten
    Cage & Aviary Birds
    Die Gefiederte Welt
    Die Voliere
    Gefiederter Freund
    Onze Parkieten
    Onze Vogels
    Parkieten Sociëteit
    Parrot Society

REGISTER

Accomodation	24
Administration	38
Aratinga acuticaudata acuticaudata	45
Aratinga acuticaudata haemorrhous	45
Aratinga acuticaudata neoxena	45
Aratinga acuticaudata neumanni	45
Aratinga aurea aurea	145
Aratinga aurea major	145
Aratinga auricapilla auricapilla	97
Aratinga auricapilla aurifrons	97
Aratinga cactorum cactorum	141
Aratinga cactorum caixana	141
Aratinga canicularis canicularis	123
Aratinga canicularis clarae	123
Aratinga canicularis eburnirostrum	123
Aratinga chloroptera chloroptera	87
Aratinga chloroptera maugei	87
Aratinga erythrogenys	77
Aratinga euops	93
Aratinga finschi	61
Aratinga guarouba	51
Aratinga holochlora brevipes	55
Aratinga holochlora brewsteri	55
Aratinga holochlora holochlora	55
Aratinga holochlora rubritorquis	55
Aratinga holochlora strenua	55
Aratinga jandaya	103
Aratinga leucophthalmus callogenys	83
Aratinga leucophthalmus leucophthalmus	83
Aratinga leucophthalmus nicefori	83
Aratinga leucophthalmus propinquus	83
Aratinga mitrata alticola	71
Aratinga mitrata mitrata	71
Aratinga nana astec	119
Aratinga nana nana	119
Aratinga nana vicinalis	119
Aratinga pertinax aeruginosa	131
Aratinga pertinax arubensis	131
Aratinga pertinax chrysogenys	131
Aratinga pertinax chrysophrys	131
Aratinga pertinax griseipecta	131
Aratinga pertinax lehmanni	131

Aratinga pertinax margaritensis	131
Aratinga pertinax ocularis	131
Aratinga pertinax paraensis	131
Aratinga pertinax pertinax	131
Aratinga pertinax surinama	131
Aratinga pertinax tortugensis	131
Aratinga pertinax venezuelae	131
Aratinga pertinax xanthogenia	131
Aratinga solstitialis	107
Aratinga wagleri frontata	65
Aratinga wagleri minor	65
Aratinga wagleri transilis	65
Aratinga wagleri wagleri	65
Aratinga weddellii	113
Aruba Brown-throated Conure	131
Aztec Conure	119
Blue-crowned Conure	45
Bolivia Conure	45
Bonaire Brown-throated Conure	131
Breeding	30
Brewster's Green Conure	55
Brown-eared Conure	131
Brown-throated Conure	131
Cactus Conure	141
Carriker's Conure	65
Chapman's Mitred Conure	71
Colombia Brown-throated Conure	131
Cordilleras Conure	65
Cuban Conure	77
Diet	27
Diseases	35
Eastern Aztec Conure	119
Ecuadorian White-eyed Conure	83
Finsch's Conure	61
Golden-cheeked Brown-throated Conure	131
Golden-crowned Conure	145
Golden-fronted Conure	97
Golden-headed Conure	97
Greater golden-crowned Conure	145
Guiana Brown-throated Conure	131
Habits	20
Hispaniolan Conure	87
Introduction	11
Jamaican Conure	119
Jendaya Conure	103
Lehmann's Brown-throated Conure	131
Margarita Conure	45
Margarita Brown-throated Conure	131

Mauge's Conure	87
Mexican Green Conure	55
Mitred Conure	71
Nicaraguan Green Conure	55
Nicéforo's White-eyed Conure	83
Pale Cactus Conure	141
Para Brown-throated Conure	131
Peter's Conure	65
Petz's Conure	123
Purchasing	17
Queen of Bavaria's Conure	51
Red-masked Conure	77
Red-throated Conure	55
Sex	20
Sharp-tailed Conure	45
Socorro Green Conure	55
Species descriptions	40
St Thomas Conure	131
Sun Conure	107
Surinam Brown-throated Conure	131
Tortuga Brown-throated Conure	131
Venezuela Brown-throated Conure	131
Wagler's Conure	65
Weddell's Conure	113
Western Mexican Petz's Conure	123
White-eyed Conure	83